PRAISE FOR *ON THE EDGE*

"On the Edge: *The State and Fate of the World's Tropical Rainforests* is the definitive assessment of the single most important factor in the future of Earth's biodiversity."

EDWARD O. WILSON, Pulitzer Prize-winning author and university research professor emeritus, Harvard University

"Tropical rainforests are, objectively, the world's habitats with by far the greatest biological richness. They are also, in my subjective opinion, the world's most magically beautiful habitats. In this new book the distinguished biologist Claude Martin summarizes the bad news and the good news about rainforest conservation, the leading threats to the world's rainforests, and the ways in which we can best deal with those threats."

JARED DIAMOND, professor of geography at the University of California (Los Angeles), and Pulitzer Prize-winning author of bestselling books including *Guns, Germs, and Steel, Collapse,* and *The World Until Yesterday*

"Of all invasive species, none has been more destructive than us. From our birth in Africa, humanity has moved across the planet and now with exploding numbers, technological power and economic demand, we are destroying ecosystems and driving other species to extinction at a terrifying rate. Life's resilience and adaptability through enormous changes over 3.8 billion years has been diversity at the gene, species, and ecosystem levels. Martin provides us with a description of the catastrophic effects of human activity and a description of some of the possible avenues away from this destructive path."

DAVID SUZUKI, internationally renowned environmentalist, recipient of UNESCO's Kalinga Prize for the Popularization of Science and the 2009 Right Livelihood Award

"The world's tropical rainforests are exposed to greater risks than ever before, now even further exacerbated by climate change. It is essential for the global community to identify solutions and take action to keep them safe. Claude Martin has a long experience in tropical forest conservation, both as a scientist and as former director general of WWF International, and in this book he presents a vitally important agenda for the 21st century to save the biodiversity and indigenous communities of the world's precious tropical rainforests. I urge you to read it."

YOLANDA KAKABADSE, president, WWF International

"To save civilization, there is nothing more urgent today than to regenerate and conserve our highly threatened forests—most importantly our tropical forests. This book comes just in time, using the field studies of dedicated scientists to show how urgent the problem is and taking examples from the efforts of committed practitioners to show how systems-based solutions could yet bring humanity back from the brink of self-destruction."

ASHOK KHOSLA, former president, IUCN and copresident, Club of Rome

"On the Edge is more than just a very comprehensive consolidation of information about the tropical rainforests: it also builds a very compelling argument about what drives deforestation and reports an array of success stories of how to conserve, regenerate, and sustainably manage this Earth treasure."

TASSO AZEVEDO, former chief of the Brazilian Forest Service and cochair of Megaflorestais

"The rainforests of the tropics now feature prominently in our collective imagination. But do we really know what has happened since the early documentaries 40 years ago? We need to know, for therein lies the hope, and Claude Martin's deeply impressive account of the state and fate of the rainforests is the place to turn while there is still a bit of time left."

JAMES GUSTAVE SPETH, author of The Bridge at the Edge of the World: Capitalism, the Environment, and Crossing from Crisis to Sustainability

ON THE EDGE

CLAUDE MARTIN

foreword by THOMAS E. LOVEJOY

ON THE EDGE

The State and Fate
of the World's
Tropical Rainforests

A Report to the Club of Rome

David Suzuki Foundation

GREYSTONE BOOKS
Vancouver/Berkeley

Greystone Books Ltd.
www.greystonebooks.com

David Suzuki Foundation
219-2211 West 4th Avenue
Vancouver BC Canada V6K 4S2

Cataloguing data available from Library and Archives Canada
ISBN 978-1-77164-140-1 (pbk.)
ISBN 978-1-77164-141-8 (epub)

Editing by Eva van Emden
Jacket and text design by Nayeli Jimenez
Jacket photograph by iStockphoto.com

Interior photographs and figures by Global Forest Watch,
M. C. Hansen/UMD/Google/USGS/NASA, Max Hurdebourcq, Zig Koch/
WWF Brazil, Claude Martin, Gleilson Miranda/FUNAI, National
Academy of Sciences US, WWF Indonesia, WWF Living Amazon
Initiative, WWF Malaysia/Mazidi Abd Ghana, WWF US

Wood engravings (pages xiv, 38, 64, 176) from Henry Emery,
La Vie Végétale: Histoire des Plantes à l'usage des gens du monde (1878), Paris :
Librairie Hachette; (page xxii) from Hermann Wagner (Ed.), Die neuesten
Entdeckungsreisen an der Westküste Afrikas (1863), Leipzig: Spamer.

Printed and bound in Canada by Friesens
Distributed in the U.S. by Publishers Group West

We gratefully acknowledge the financial support of the Canada Council
for the Arts, the British Columbia Arts Council, the Province of British
Columbia through the Book Publishing Tax Credit, and the Government of
Canada through the Canada Book Fund for our publishing activities.

Greystone Books is committed to reducing the consumption of old-growth
forests in the books it publishes. This book is one step toward that goal.

A MESSAGE FROM THE CLUB OF ROME

ON THE EDGE: *The State and Fate of the World's Tropical Rainforests* is the thirty-fourth Report to the Club of Rome since *The Limits to Growth*, the first Report to the Club of Rome, was published in 1972. It has been peer–reviewed by our experts to ensure it is scientifically rigorous and contributes vital new elements to an important global debate.

The Club of Rome was one of the first and, for many years, one of the most influential thinkers on long-term economic, social, and environmental issues. It kick-started the environmental movement, has influenced the thinking of millions of people for more than a generation, and has shaped enlightened policy decisions around the world. It placed long-term thinking about humanity's most important challenges at the top of the international agenda.

With *On the Edge*, Claude Martin presents the history and current state of knowledge on the loss of the world's tropical rainforests and their role for the global climate. It is a wake-up call, meant to help everyone understand the threats and global consequences of continued destruction. It also highlights more recent positive developments and outlines the measures needed to ensure future generations will also be able to enjoy the magic of the millions of square kilometers of untouched rainforest that still remain.

The Club of Rome believes that *On the Edge* provides a crucial contribution for us all to better understand and address the root causes of deforestation and forest degradation. It presents the principles of tropical forest conservation and explores the options for saving a large part of the world's biodiversity and the cultures of indigenous communities.

On the Edge puts tropical rainforests back where they belong: at the center of the environmental debate.

CONTENTS

FOREWORD

W HEN P. W. Richards wrote *The Tropical Rain Forest* in 1952, these forests were distant and exotic to most of the world of science. There was no remote sensing to give the latest information on deforestation and the extent of the forest. Largely ruled by colonial powers, the rainforest was mostly inhabited by indigenous and other local peoples. While a tiny number of scientific institutions performed research in tropical forest regions, science consisted mostly of "expeditions," with all the sense of the exotic that phrase implies.

Today, of course, there is almost minute-by-minute satellite coverage, and lidar has emerged with the power to identify individual tree species in what is probably the most biologically diverse habitat on Earth. And the forest has shrunk in the face of voracious appetites for tropical resources (hardwoods, fossil fuels, and minerals), advancing agricultural frontiers (no matter how ill-advised), and mostly ill-conceived settlement initiatives.

It has been close to three-quarters of a century since Richards's book appeared, and in that time the tropical forest has shrunk to a staggering degree: much of Brazil's Atlantic forests are gone, along with 20 percent of the Amazon and most of the lowland forests of Malaysia and Indonesia. The Congo Basin has suffered multiple inroads (for mining, oil and gas, timber, roads, and dams), and much of the West African forest has been obliterated.

The picture, happily, is not entirely negative. More than half of the Amazon is now under some form of protection. Protected areas (however vulnerable) have been created in the Congo Basin, Southeast Asia, and Indonesia. National capacity in science and conservation has grown impressively.

At the same time science has revealed how important these forests are not only for their extraordinary biological riches, but also for the role they play in global cycles, including carbon and water. The Amazon rainforest acts as a flywheel to moderate the continental climate, generates half the continent's rainfall, provides moisture beyond the Amazon Basin, and influences the global climate.

Tropical forests store immense quantities of carbon, which makes them highly important in managing the threat of climate change caused by excess carbon dioxide in the atmosphere. Not only is it important to recognize the sensitivity of living systems to climate change, but it is also essential to recognize that the planet works as a linked biological and physical system, and therefore we must manage it as a living planet. The enormous dividend, if we succeed in doing so, will be conserving the remaining biodiversity and its potential for the benefit of future generations.

So the time is indeed at hand for a book such as this: one that goes beyond the science of the tropical wonderland, and considers the complex challenges, pressures, and solutions. My colleague from my World Wildlife Fund days, Claude Martin, has recruited some of the very best minds working on different aspects of managing and conserving these incredible forests. The result is a volume of quality and relevance very deserving of succeeding Richards's classic volume, and it should ensure there will be thriving tropical rainforests for another three-quarters of a century hence.

THOMAS E. LOVEJOY

Professor of environmental science and policy, George Mason University, Fairfax, Virginia, former assistant secretary for environmental and external affairs, Smithsonian Institution, and chief biodiversity advisor, World Bank.

ACKNOWLEDGMENTS

THE SUPPORT AND encouragement from the staff of the Club of Rome secretariat, particularly of Karl Wagner and Alexander Stefes, have been essential in writing this book. I would also like to thank Ian Johnson and Thomas Schauer, as well as the Club of Rome members who reviewed the manuscript. I have greatly benefited from the advice on system dynamics provided by my friend and colleague from WWF, Jørgen Randers, who has pioneered the thinking on global limits ever since he coauthored *The Limits to Growth* in 1972. I am immensely grateful to the tropical forest experts who contributed to the content of the book with a "Specialist's View" section or a box: Jürgen Blaser, Bruce J. Cabarle, Chris Elliott, David Kaimowitz, Cláudio C. Maretti, Ralph M. Ridder, Rodney Taylor, and Jeffrey Sayer. I am particularly indebted to Thomas E. Lovejoy, a world-renowned authority on tropical forest conservation, who contributed the foreword, and to Matthew C. Hansen from the University of Maryland, who provided essential guidance on complex remote sensing interpretations. A number of current and former scientific and professional colleagues also helped with cartography and advice on specific issues: Adam Dixon, Chris Hails, Max Hurdebourcq, Simon Lewis, Richard McLellan, Mathias Nagel, Stephan Wulffraat, and the experts from WWF's Living Amazon Initiative. My son Aurel Martin did the vector graphics of many of the black-and-white illustrations. I would also like to thank Greystone Books and my editor, Eva van Emden, who competently made my English more accessible to the reader without losing factual accuracy. Last but not least, the moral support of my wife, Judy, who endured my distant mind for months, was absolutely crucial.

INTRODUCTION

Überall geht ein frühes Ahnen dem späteren Wissen voraus
(An early premonition always paves the way to later knowledge)
ALEXANDER VON HUMBOLDT

I WAS LIVING IN the central Indian jungles, working as a young and rather innocent biologist, when the first Club of Rome report, *The Limits to Growth*, was published in 1972. It made a lasting impression on me, as it did on many people who were concerned about the consequences of unbridled growth and consumption of resources. This report did not deal with the issue of the limits of the world's tropical rainforests. The extent of these forests was not known at that time, and public discussion of their fate had not yet taken off. Concerns had certainly been expressed by a few scientists, but the level of knowledge about these forests had hardly gone beyond what was described in the historical records of early explorers, missionaries, and colonial forest agents. Many tropical countries, even the biggest ones, had no idea how much rainforest they held.

Since 1972 our knowledge has expanded greatly. The scientific literature on tropical rainforests has virtually exploded, particularly in the past two decades. Today we have satellite remote sensing technology that can detect an elephant in the middle of the Congo Basin and monitor forest clearings the size of a backyard. We have the tools to understand what drives tropical land use change. But will the international community use these tools to effectively address tropical deforestation and loss of biodiversity? Or will we see a gradual demise of these forests, much as we are already seeing the disastrous consequences of climate change in many parts of the world?

Scientific institutions as well as nongovernmental organizations now generate hundreds of articles and reports every year on tropical forest cover change, forest fragmentation, biodiversity, climate change, and carbon storage. Most scientists, however, work within their academic disciplines and publish their results in specialized journals, without much cross-referencing or synthesis. Research papers are highly technical and difficult for nonspecialists to read. Unless one of the more broadly oriented journals such as *Science* or *Nature* publishes research findings with wide enough appeal to be picked up by the media, important scientific information may never become available to policy makers and the public.

Language barriers make this problem still worse. Today most scientific literature about tropical rainforests comes from universities in the United States and the United Kingdom, and science institutions in non-anglophone countries often have difficulty following the newest research. But the inverse is also true. Research on rainforest ecology of the former French and Belgian colonies in Africa rarely appears on the radar of anglophone scientists. This is one reason for the common belief that African rainforests are biologically impoverished and not well understood. It may be only the language of the studies that is not well understood. In a wonderful book by

Jean-Pierre Vande weghe, *Forêts d'Afrique Centrale, la Nature et l'Homme*, about half of the more than 350 literature citations are to French publications that deal with the biology of African rainforests.

The mysteries and the beauty of the world's jungles still attract great curiosity and stir deep emotion among a very large audience. No other vegetation zone has generated as many books and films as the tropical rainforest. Are these not the places where our closest relatives still survive? Yet since the start of the new millennium, public concern about tropical rainforests seems to have waned. With the rising number of global environmental problems, some conservation organizations have also started to suffer from "rainforest fatigue." As the world has become preoccupied with climate change, people have begun to see tropical rainforests simply as carbon sinks. This grossly reductionist view has redirected public attention and concern for the survival of the tropical rainforest to a technical and academic level of discourse that is uninteresting or out of reach for most people.

In the past few years I have increasingly found that many people believe tropical rainforest destruction is not an issue anymore or that the cause is hopeless. Both beliefs are untrue, but without public attention there will not be enough political pressure to counteract the immense commercial forces that can destroy the remaining areas of untouched tropical rainforest.

In search of the unvarnished truth

IN THIS BOOK I have tried to create an overview of the current state of knowledge about the world's rainforests. I wrote this book not just because I have always been fascinated by tropical rainforests, but also because I wanted to discover the unvarnished truth about why we are losing these forests that are so vital to the future of our planet's biodiversity, and what we can do to save at least some of them. I hope that by sharing my findings, I can attract the interest and prompt the motivation of other people as well.

Although I have tried to cover the most important findings from the scientific literature, particularly the most recent work, this book discusses only a fraction of the thousands of articles and books published in the last few decades. A book like this can obviously never hope to cover the huge wealth of knowledge about the biodiversity of all the tropical rainforests and the indigenous communities that live in them. I have only dealt with those aspects and trends that I consider most relevant for the future of the majority of tropical rainforests. To round out my judgment and experience, I have asked eight tropical forest experts to contribute their own views on specific issues.

I think it is important to take a pragmatic view of the issues affecting the future of tropical rainforests. The Food and Agriculture Organization of the United Nations projects a 60 percent increase in the world demand for agricultural products between 2015 and 2050, a consequence of human population increase, urbanization, and dietary changes, particularly the rapid increase of meat and palm oil consumption in emerging economies. The world may literally eat up the tropical rainforests, these being the only large areas still available for agricultural expansion. Unfortunately, it is more profitable to produce a few tons of beef or animal feed per hectare than to maintain that hectare as intact tropical rainforest. As long as our economic models lead to such drastic failures, national governments and the international community, to say nothing of the indigenous forest dwellers, remain weak, if not totally powerless.

Romanticizing the tropical rainforest certainly does not help us understand and preserve it, and I will stay clear of any such notions, especially as many others have done this more convincingly than I ever could myself. This is not to say that I do not feel a deep admiration for the early descriptions of the wonders of the tropical rainforest by people like Charles Darwin, Alexander von Humboldt, and P. W. Richards. I have strong memories of many moments of joy and amazement when I lived in the tropical rainforest: listening to the deep reverberations of chimpanzees drumming against large tree buttresses to mark their territories, or watching from the terrace of my bungalow in western Ghana as the carmine sun turned deep purple before going down over an unbroken canopy of rainforest trees. I remember my astonishment when big fish belonging to half a dozen different species jumped into our boat as we drove slowly up a narrow tributary of the Amazon River.

But I have also seen the less delightful realities of these forests when I got lost in remote areas, accompanied only by a local hunter with his machete. I have also walked into tiny forest settlements where every inhabitant was suffering from elephantiasis—swollen legs caused by a parasitic worm. Tropical rainforests can be threatening for those who live in them. At those times they are far removed from the orchid-bedecked wonderlands of Walt Disney's *Jungle Book*—sometimes the jungle is just hot, green, oppressive, and full of driver ants.

No place for pessimism

I HAVE OFTEN been asked after my speeches on tropical rainforest conservation whether I am an optimist or a pessimist. My standard answer is that this depends on whether I had breakfast or not. My evasiveness is rooted in uncertainty about the economic, social, and political factors that will determine the future of these forests. I doubt that any honest person could foretell the future of the rainforests to the end of the twenty-first century. I have felt intense pessimism when I saw huge expanses of rainforests destroyed with bulldozers for cattle pasture, or when some of the richest lowland rainforests in Kalimantan, Indonesia, were turned into yet another palm oil plantation. On the other hand, I felt some optimism when I learned that a practically unexplored area of rainforest the size of the Netherlands, the Montanhas do Tumucumaque in the Brazilian state of Amapá, bordering French Guiana, had been declared a protected area (see plate 18). And I looked with optimism at some of the amazing pictures of indigenous people from an uncontacted tribe pointing their arrows at a plane from the Brazilian National Indian Foundation (FUNAI). Sixty-seven uncontacted tribes have been identified in Amazonian indigenous reservations (see plate 16).

We have to consider both of these realities—the damage likely to be done in the future as well as the chances for rainforest preservation—for our projections about the future to have any degree of credibility. Not long ago I heard a well-known French botanist deplore the fate of the tropical rainforests. He claimed that by now only degraded secondary forest remained. I certainly do not want to downplay the tremendous pressure on the rainforests and the very real threat of massive further deforestation, but such blatantly wrong throwaway remarks are not only misleading, they also undermine support for urgently needed rainforest protection measures.

When the great scholar and explorer Alexander von Humboldt embarked on his five-year expedition to Central and South America in 1799, accompanied by the botanist Aimé Bonpland, no European had penetrated the deep Amazonian forests before them. On their journey on foot and by boat they proved that the mysterious Rio Casiquiare indeed connects the Orinoco River with the Amazon Basin, a phenomenon that European geographers of the time believed to be impossible. The indigenous people knew about the Rio Casiquiare, just as they had names for all the plants and animals that overwhelmed Humboldt and Bonpland to the point where they were afraid they would be driven mad by the wealth of new species. Most of the rainforests of the world were, and still are, inhabited by people, but at the time of Humboldt's journey the world population was only one billion, less than one-seventh of today's population. The scars from the indigenous forest dwellers who lived from shifting cultivation were quickly healed over again by the forest. The area covered by tropical rainforests at that time was still close to the 1.6 billion hectares believed to be the original maximum distribution area. And with the exception of limited areas under shifting cultivation, these forests were essentially primary forests—forests that had never undergone any significant disturbance. Today less than half this area remains as undisturbed forest—nobody knows exactly how much—and about another quarter survives as fragmented and degraded forest.

The oil palm (*Elaeis guineensis*), known and used by the local people of West Africa for centuries, is now grown in huge plantations, mostly at the expense of intact rainforests.

1

A DESTRUCTIVE TWENTIETH CENTURY
The Rising Consciousness

IT IS NOT a coincidence that people became aware at the end of the
1960s and early 1970s that tropical rainforests are not an unlimited resource, at the time when the *Limits to Growth* report to
the Club of Rome[1] was in preparation. Until the economic boom
after World War II, tropical forests were believed to be "exuberant,
resilient and indestructible."[2] Some timid forewarnings in earlier
publications had passed without the public, let alone politicians
and economists, taking any notice. It wasn't until a few decades
later that the paradigm of limitless growth started to be questioned,
at least by a few academics. Although some of the modeling in *The
Limits to Growth* may not have been accurate, the report sent a timely
message that we must break with the outrageous assumption
that humanity could continue to consume the planet's resources
indiscriminately.

Tropical clearing in earlier centuries

WHEN PEOPLE VOICED their concerns over the fate of tropical rainforests in the postwar years, they made little reference to tropical areas having been deforested before. In fact, the majority of the tropical rainforests had been inhabited by people for a long time, although at very low population densities. In many tropical rainforest areas, even in forests considered to be "primary," or undisturbed, one can find traces of former human habitation and shifting agriculture. Thus "primary" does not mean that there has never been any human disturbance. Not much is known about tropical deforestation rates in historical times, but it seems likely that between 1700 and 1850 less than one million hectares was cleared annually, almost exclusively for shifting agriculture.[3] And since at the same time the forest was regrowing in the small abandoned clearings, this rate of deforestation had hardly any effect on the overall extent of the tropical rainforest.

Michael Williams, the author of *Deforesting the Earth*, describes how deforestation accelerated in the colonial period after 1850.[4] Rice cultivation started to take a heavy toll when immigrant peasant farmers cleared large areas in the lowland forests of Myanmar, Thailand, Cambodia, Laos, and Vietnam. These countries became an important rice exporting region—the rice bowl of Asia. Other Southeast Asian countries followed, deforesting large areas of Sumatra and Java. Over ten million hectares of forest was cleared before 1920, although records of how this happened seem to be inconclusive. The Philippines, which had been under more than 300 years of Spanish colonial rule, also lost forest in the central planes of Luzon, mainly to plantations of commercial crops such as tobacco and sugar cane. The colonial influence on the expansion of commercial crops was also noticeable in Assam, India, where the British East India Company expanded its tea estates at the cost of tropical forests. In West Africa, mainly in Ghana and Nigeria, rainforests were cleared in the last quarter of the nineteenth century for some of the first commercial cocoa and oil palm plantations.[5]

The rainforests of Latin America, on the other hand, had hardly been affected by commercial agriculture before 1920, with the exception of some coffee plantations, for example in the areas northwest of São Paulo. Most of the deforestation in South and Central America was caused by shifting agriculture, which usually affected relatively small areas compared to cultivation in South and Southeast Asia. A notable exception to this pattern of moderate deforestation was the Brazilian Atlantic coastal forests, which had already been decimated in the initial phases of colonization in the Portuguese-controlled areas.[6]

FASTER DEFORESTATION FROM 1920 ONWARD

AT THE GLOBAL level, the level of deforestation remained modest until about 1920, as agricultural expansion happened mainly in temperate countries until that time.[7] The world population had grown from 1 billion in 1800 to 1.9 billion by 1920. The majority of the population growth had taken place in the industrializing countries of the West, as well as South Asia.

What happened in the following eighty years up to the end of the twentieth century, during a period when the world population more than tripled, left little hope for the future. Tropical rainforests became the target of the industrializing world's rapidly increasing demand for timber as well as for agricultural expansion for the production of crops and meat. Through the 1920s and 1930s, tropical deforestation accelerated massively to almost eight million hectares per year.[8] As no forest inventories existed in most tropical countries, this is a very rough estimate, and there is not much information available on how and where this deforestation actually happened.

In the aftermath of World War II, from 1950 to 1980, tropical deforestation increased even further. An estimated eleven million hectares of tropical rainforest disappeared every year in the 1970s—at least, this was the widely cited figure from a first assessment published by the Food and Agriculture Organization of the United Nations (FAO) in 1976. This study concerned the period that Michael Williams called the "Great Onslaught," when the demand for timber, cropland, and pastures shifted from the industrialized countries to the tropics. After political independence, former colonies did not necessarily also become economically independent. The voracious demand for timber, coffee, cocoa, sugar, and other agricultural products in the postwar period and the economic boom years of the 1960s continued to put a heavy load on the world's rainforest areas. Meanwhile, deforestation slowed in the temperate zones of the Western countries, many of which experienced "forest transitions"—a gradual return of their forest areas resulting from changing land use patterns (see chapter 5). In addition to the demand from the former colonial powers came the rapidly increasing imports by Japan, which became the largest tropical timber importer in the world in 1960. The vast majority of the timber imported by Japan originated in the less than sustainably exploited forests of the two Malaysian states of Sarawak and Sabah.[9]

The world wakes up

IN THE ABSENCE of official statistical information on forest degradation and the conversion of tropical forests to other land uses, the first alarm calls came from scientists. A number of botanists were particularly apprehensive now that machetes and axes had been replaced by bulldozers and chain saws, allowing much faster clearing of tropical forests. At the time, their concern was less about the environmental, cultural, and economic effects of losing an entire biome than that their study subjects—as yet unknown plants—might disappear before they could be studied. A mycologist at the

Singapore Botanic Gardens expressed it in these terms: "I fear lest all virgin lowland forest of the tropics may be destroyed before Botany awakes; even our children may never see the objects of our delight which we have not cared for in their vanishing."[10]

The most often cited author to provide a vivid appreciation of the richness and tremendous value of tropical rainforests was P. W. Richards, an English botanist who published the popular science book *The Tropical Rain Forest* in 1952:[11] "Whole chapters of biology may never be written," he predicted, if the destruction of rainforests continued at the same rate as in the first years after World War II. Richards was one of the rare botanists who had intensive field experience in all three of the world's major tropical regions. He was certainly an authority on the "wonderfully varied life" of these forests—but who would pay attention to the admonitions of a single botanist in a postwar world that needed timber more than warning bells from scientists?

Some of the people driving the awakening concern about the fate of the tropical rainforests were scientists who had seen the consequences of increasing exploitation of the seemingly unlimited forest in their own countries. One of the most prominent was Arturo Gómez-Pompa, a Mexican botanist who became a professor at the University of California, Riverside. In the same year that *The Limits to Growth* appeared, he published a much-cited article in the journal *Science*, together with some of his students: "The Tropical Rainforest: A Nonrenewable Resource." It was based on a full-scale ecological survey Gómez-Pompa had made of the Mexican rainforests.[12] He had observed that tropical rainforests have a different regeneration system than other forest types. It functions well to close small gaps left by tree falls, storms, or flood damage because remaining tree seedlings and the seed stock of forest trees in the soil guarantee a relatively quick recovery of the understory. This capacity of tropical rainforests to close small gaps explains why shifting cultivation by indigenous peoples at low densities did not lead to lasting forest damage. However, Gómez-Pompa's work left no doubt that the land use practices that became common after World War II would exceed the regenerative capacity of rainforests. Larger clearings would change the soil and microclimatic conditions to such an extent that the seedlings and dormant seeds of closed forest trees would not be able to survive. This finding might seem obvious to us now, but it is as important as ever to understand that tropical rainforests cannot regenerate the same way that we see temperate forests expanding again on the deep soils of the northern hemisphere.

DISTRESS SIGNALS FROM SCIENTISTS

SHORTLY AFTER GÓMEZ-POMPA'S article, the geographer William M. Denevan wrote an article that shocked a great many people. It was titled "Development and the Imminent Demise of the Amazon Rain Forest."[13] Denevan, who had seen the Amazon increasingly divided by roads, converted to cattle ranches, and used for mining, predicted, "Within one hundred years, probably less, the Amazon rain forest will have ceased to exist." This was certainly an alarmist statement without statistical evidence, but objective information simply did not exist at that time. Denevan had crossed the entire Amazon Basin and was not writing as an armchair scientist. In fact, he foretold some of the large-scale developments in the Amazon in the following decades and triggered worldwide concern expressed through hundreds of news articles in the late 1970s and 1980s. Denevan became a much-revered professor of geography at the University of Wisconsin, Madison. His main work was devoted to the aboriginal cultures of South America. His article "The Pristine Myth: The Landscape of the Americas in 1492,"[14] which asserted that the native peoples of the Americas changed the landscape, attracted at least as much attention and controversy as his prediction about the future of the Amazon. Many scientists were disturbed that he was contradicting historical records, including those of Christopher Columbus, who had described the Americas as a "paradise of untouched primeval forests."

We can credit a few courageous scientists with making tropical rainforest destruction a major public concern in the 1970s. Not only did they break away from the paradigm of the limitless and resilient resource, they also dared to make projections without adequate statistical information. It was important that these scientists published their views not as scientific papers but as articles and popular books. This allowed them to reach the general public and lent academic credibility to the issue of tropical deforestation, showing that it was not just a concern for a bunch of emotional naturalists or alarmist nongovernmental organizations crying wolf. Tropical deforestation would soon become an issue of global dimensions and large, credible international nongovernmental organizations soon started to become active.

First systematic attempts to assess the tropical rainforests

AS THE ALERTS of the early 1970s were increasingly taken up by the media, the FAO found itself in a somewhat awkward situation. The Division of Forestry and Forest Products of the FAO had come into existence in May 1946 (the FAO was founded in 1945), and two international conferences on forest statistics had taken place. However, the FAO categorization of forest into "productive" and "other forest" was obviously not useful for quantifying the loss of tropical rainforest. The FAO's attention in the first decades after World War II was primarily focused on increasing timber stocks and productivity, and it did not make a distinction between subtropical and tropical dry or moist forests.

In 1968 the Swedish forester Reidar Persson joined the FAO
Forestry Department. He started to analyze the FAO's country questionnaires on forest resources, and then made study trips to tropical
countries in Asia, Africa, and Latin America. He was interested in
finding a new approach to the World Forest Inventory, a survey
that the FAO's Forestry and Forest Products Division had carried
out four times. These inventories had little value. The information
was gathered with constantly changing questionnaires and different
interpretations of what constituted a forest. Some countries inven-
toried as little as 15 percent of their forest cover, and others did not
complete the questionnaires at all. Persson estimated that more
than half the data from tropical countries were less than 40 percent
accurate.[15]

A further problem was that the objective of the World Forest
Inventory was as vague as the results. The World Forest Inventory
attempted to provide a general picture of the extent of the world's
forests, but it focused mainly on supplying wood to meet future
demand. It did not consider secondary forest products or other ecosystem services such as the forest's role in preserving water cycles
and biodiversity.

The situation changed in the mid-1970s when the Swiss government assigned a young forester, Adrian Sommer, to the FAO to take
on the difficult task of making the first systematic assessment of the
world's tropical moist forests.[16] During this period the term "tropical moist forest" was used as a substitute for a more comprehensive
interpretation of "tropical rainforest" (see box 1.1. and figure 1.1).
Sommer's assignment was obviously motivated by rising concern
among a broad segment of the Western public and the fact that the
United Nations organization responsible for the monitoring of the
world's forest resources had not been in a position to provide any
information on the rising problem.

FIG. 1.1. Rainforest types of Ghana according to Hall and Swaine

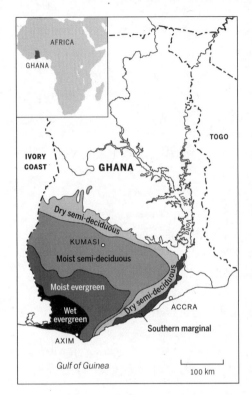

Two British botanists, John B. Hall and Michael D. Swaine, classified the rainforest area of Ghana.[17] The forest types were defined by analyzing the composition of plant species and then named according to climatic and physiognomical criteria. A schematic cross-section through the rainforest zone in Ghana from the coast near Axim (left side of the drawing) moving north shows the four rainforest types: wet evergreen (more than 1,750 millimeters rainfall), moist evergreen (1,500–1,750 millimeters), moist semi-deciduous (1,250–1,750 millimeters), and dry semi-deciduous (1,250–1,500 millimeters). The tallest trees (more than 50 meters) occur in the moist semi-deciduous zone, which is the main timber harvesting area. Illustrations from Martin.[18]

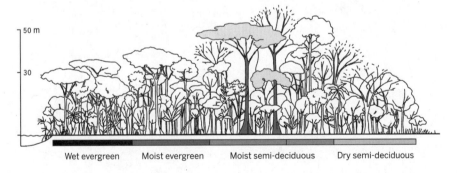

Wet evergreen   Moist evergreen   Moist semi-deciduous   Dry semi-deciduous

BOX 1.1 ## "Tropical rainforests," "tropical moist forests," or "humid tropical forests": Going in circles

The term "rainforest" goes back to the term "tropischer Regenwald" first used by the German botanist A. F. W. Schimper in his *Plant-Geography upon a Physiological Basis* of 1898.[19] But it was only in 1952 that P. W. Richards, in his momentous book *The Tropical Rain Forest*,[20] made the term commonly known. He described what he meant in a remarkably precise manner. Tropical rainforests occur on all three tropical continents as well as the Pacific Islands, roughly between the Tropics of Cancer and Capricorn with a few exceptions. Rainforests extend slightly across the Tropic of Cancer into Assam in India and across the Tropic of Capricorn along the east coast of Australia.

In the introduction to his book, Richards stated that it would be desirable to construct a "biological spectrum" to classify this biome and sketched out a plant-sociological framework for it. A global ecoregion map based on such criteria was recently established by WWF US (see plate 1). Richards's insights were overlooked by some later authors who defined tropical rainforests based on rainfall thresholds of 2,500 millimeters or even more. This led to a much more restrictive definition, under which large areas in Africa and in the Amazon were classified as some form of drier seasonal forest instead of rainforest. We now consider rainforests to also occur in tropical areas with considerably lower annual rainfall, and a relatively even seasonal distribution of rainfall through the year is much more relevant than total rainfall.[21] Among other problems, the confusion led to a bewildering variety of rainforest distribution maps covered with imaginary green blotches, which one can now find on many Internet sites.

In 1976 the term "tropical moist forest" was introduced by Adrian Sommer as he attempted to assess the extent of these forests. He included tropical semi-deciduous forests (where some upper canopy trees are drought deciduous) in his definition. Norman Myers defined tropical moist forest as "evergreen or partly evergreen forests, in areas receiving not less than 100 mm of precipitation in any month for 2 out of 3 years." In more recent scientific literature, rainforests are often called "humid tropical forests" to correspond with the humid tropical ecozones and to signal that this includes the tropical ombrophilous (thriving in heavy rainfall), as well as the tropical evergreen seasonal forests (see figure 1.1). Many of the most recent scientific papers are now using the more popular term "tropical rainforest" again. Thus, we have come full circle to what Richards taught us in 1952 when he included a fold-out map of the tropical rainforest climax distribution area, which was remarkably accurate for the time.

More than sixty years later, when I use the term "tropical rainforests" in this book, it always has the broader meaning encompassed by "tropical moist forest" or "humid tropical forest" biomes, as distinct from deciduous and dry tropical forest biomes. Note that these terms leave many classification questions open, such as whether they include swamp, montane, or mangrove forests. Without a clear definition, a comparison of area estimates is impossible. Rainforests also exist in temperate zones, for example on the Pacific coast of California, but if I refer simply to "rainforests" in this book, it should always be understood to mean "tropical rainforests."

PILES OF COUNTRY FIGURES BUT VERY FEW FACTS

SOMMER AND HIS team encountered great difficulties with what they described as a "voluminous mass of confusing reports, scattered all over the world, yielding very few facts." Most tropical countries at that time had not completed comprehensive forest surveys, or their forest inventories covered only accessible areas near coasts and large rivers. Another major problem was differences in forest classification, especially in distinguishing forest from wooded savanna. Sommer chose the vegetation classification system established by UNESCO in 1973.[22] For his assessment he included fourteen forest types to define the overall area of the world's tropical rainforests. These included all the evergreen ombrophilous (rain-loving) forest types—the rainforests in the strictest sense of the term—but also South Asian semi-deciduous forest types as well as submontane and montane forests (cloud forests). He then compared their historical maximum distribution areas with the actual areas covered.

The question "What is a forest?" may sound trivial, but forest classification is one of the greatest obstacles to measuring forest cover change. A document on the Forest Information Services website,[23] which is regularly updated, lists more than 1,500 definitions of forest and forest land, of which almost one hundred are internationally used definitions for forest as land cover type. Only recently did the FAO and other international forestry organizations try to harmonize definitions. This effort was triggered by the need to reconcile forest definitions among the parties of the United Nations climate change convention, which is concerned with carbon storage and changes in land use.[24]

Taking into account all the uncertainties in the survey data, Sommer nonetheless compared a supposed global historical maximum tropical rainforest area of 1.6 billion hectares with the estimated area reported in the first half of the 1970s (see appendix 1). This was only meant to give a very rough indication of the forest area lost since historic times. Subsequent assessments using remote sensing data have since revealed that Sommer actually underestimated the tropical rainforest cover level in the 1970s. The decades after World War II were undoubtedly marked by accelerated deforestation in tropical forests, but even though it was obvious that remote forest areas were opening up and forest was being converted to commercial farmland, Sommer did not have enough data to calculate deforestation rates. He nevertheless ventured a guess that about eleven million hectares of tropical moist forests were lost each year worldwide. He based this figure on the deforestation rates in only thirteen of the sixty-six countries examined. In hindsight Sommer must have regretted his bold estimate; despite his disclaimers, it was quoted as fact and often translated by the media, nongovernmental organizations, and researchers into football fields cleared per minute.

In this first assessment Sommer expressed hopes that an FAO monitoring program using satellite imagery would soon provide an accurate appraisal of tropical forest cover. The first imagery from NASA Landsat satellites had already become available in 1972–73. Today, forty years later, we still do not have an officially recognized global forest cover monitoring program that could provide a comprehensive assessment of the tropical forest cover. University departments and nongovernmental organizations, such as the World Resources Institute, are partly filling this role. The reasons why we now receive high-resolution pictures from the surface of Mars but do not know where different forest types remain on our own planet will be examined in the following chapters.

A desperate lack of data from the largest rainforest countries

IN 1978 THE FAO and the United Nations Environment Programme (UNEP) started working on the Tropical Forest Resources Assessment Project, using existing documentation supplemented by

Landsat imagery.[25] Like all FAO assessments, it depended heavily on individual countries being willing and able to make information available. Both Persson and Sommer had pointed out obstacles to accurate reporting in certain countries: technical shortcomings, lack of human capacity, and reluctance to release politically sensitive information.

In 1980, even before the FAO/UNEP assessment was completed, the National Research Council of the United States National Academy of Sciences published a report by Norman Myers on the conversion of tropical moist forests.[26] It was a reaction to the lack of reliable information on the extent of tropical forests and the rate of deforestation. As Myers observed, reports published by academia and the government dealt with all kinds of scientific aspects of tropical forests, but hardly took the trouble to analyze the extent of the forest cover at a country level. In less than a year, Myers had collected, collated, and evaluated many bits and pieces of information on the status and conversion of tropical rainforests from over forty countries (seventy countries around the world fall within the humid tropical forests biome, and of those, about forty have or formerly had large forest areas).

Norman Myers is a prolific worker who published a number of books on biodiversity and species extinction around this period,[27] and he had access to other sources of information than the official government channels of the FAO in many of these countries. But his remarkable desk study ran into the same difficulties as the FAO assessments: his estimate of the forest cover of the Democratic Republic of the Congo (then Zaire) used data from 1972 with a margin of error of 40 percent! The data from Indonesia were at least twenty years old and did not account for the widespread logging and forest conversion that had taken place in the meantime, and Brazil had not yet published any statistical information on the largest area of tropical forest of any country!

Despite the desperate lack of reliable information, Myers did come across some solid evidence. At the end of the 1970s, the Philippines and Thailand published the results of comprehensive forest surveys that used remote sensing data. These surveys revealed that the Philippines had only 38 percent forest cover and not 57 percent as estimated about ten years before, and Thailand had only 25 percent and not (or no longer) the 48 percent estimated at the beginning of the decade. Although these estimates from the Philippines and Thailand could not be extrapolated to the rest of the world's tropical rainforest, this information suggested the possible, or even likely, fate of many other forest areas. Myers guessed that slash-and-burn farming alone could account for a loss of more than 1 percent per year, lending some credence to Sommer's estimated loss of eleven million hectares per year.

Two years earlier, at the Eighth World Forestry Congress of 1978, the director general of the FAO had suggested that the loss could be closer to twenty-one million hectares of tropical moist forest per year. Given that Myers had corroborated Sommer's estimate, the FAO director general's estimate was clearly an alarmist figure that lacked any foundation.

AN INCONCLUSIVE FAO/UNEP TROPICAL FOREST RESOURCES REPORT

WHEN THE TROPICAL Forest Resources Assessment Project was completed after three years, in mid-1981, both the scientific world and the broader public hoped they would finally learn what was happening to the rainforests of the world. The project produced an impressive mass of data, summarized in four technical reports. It classified the natural tropical forests into two main groups: closed forests and forest-grassland tree formations. These two groups were then classified as unmanaged and managed, productive and unproductive, forest fallow (woody vegetation developing in abandoned clearings in areas of shifting agriculture) and annually deforested areas. The reports considered all woody vegetation with a ground cover of at least 10 percent and included coniferous and bamboo forests in the various management and production classes as well. Forest plantations were classified separately.

But because some countries had only very vague or outdated ideas of their forest areas, these subdivisions became hypothetical. Another difficulty was the lack of data on the various types of closed forests. Even though the reporting countries described the forest types in their countries, and the FAO applied the UNESCO International Classification and Mapping of Vegetation System[28] to make these descriptions compatible, most countries had no numerical data on the areas concerned. It was therefore impossible to tell how much of the closed forest area in a country was tropical moist forest and how much was dry forest.

TROPICAL FORESTS SEEN AS A SOURCE OF TIMBER

THE FAO/UNEP TROPICAL Forest Resources Report of 1981–82 had a strong focus on economic productivity. This was ironic because the "productive" forest areas were being converted to other land uses most rapidly, so that the bulk of the productivity information quickly became obsolete. This was particularly true in Southeast Asian countries, which had lost half or more of their forests in a few decades and continued to lose forest area rapidly. Because forest destruction was happening so quickly, these countries tended to report how much declared forest land they had, irrespective of whether it actually contained forest or not. The combination of these factors—inaccurate or outdated information in country briefs, confusion between declared versus actual forest area, and the lack of distinction between moist and dry forest types—made the FAO/UNEP report of 1981–82 inconclusive. Although it attempted to complement Sommer's initial assessment, it actually did more to blur than to clarify the situation.

Underestimation of rainforest area in the 1970s

IF WE COMPARE these early assessments with the far more accurate satellite data of later years, we see that in 1980, forested area was underestimated by at least 10 percent. If the annual loss during this period was indeed about eleven million hectares per year, a backward projection from the more reliable 1990 figures suggests that the tropical rainforest area in 1980 was closer to 1,270 million hectares, not 1,125 million hectares as the FAO/UNEP assessment suggested (see appendix 1). The major reasons for the underestimate were lack of data for the Amazon and Congo Basins, as satellite imagery of later years revealed,[29] and underestimation of the forested area of Southeast Asia, which is spread out over more than 13,600 islands in Indonesia alone.

The estimates of tropical deforestation before 1970 were even less certain. By the most commonly cited estimates, 235 million hectares had disappeared between 1920 and 1950, and another 318 million hectares vanished between 1950 and 1980.[30] The estimated total of 553 million hectares (more than the combined areas of India and Indonesia) lost in those sixty years no doubt included areas of dry forest types cleared in South Asia and continental Southeast Asian countries during this period. Subtracting these areas would explain the gap between the presumed historical maximum forest area of 1,600 million hectares as it may still have existed some two centuries ago and the approximately 1,270 million hectares that remained in 1980.

The great uncertainties of deforestation estimates before 1980 suggest, interestingly, that the worldwide concern about tropical rainforest destruction in the 1970s had no solid numerical basis. We do not really know how much tropical rainforest was lost, and where, in the 1970s or whether Sommer's guess of eleven million hectares of annual deforestation was close to reality or not. He was the first to admit that his figure was just a guess, and there it will rest—forever. Classifying and assessing forest areas on a regional and global scale was too difficult for the methodologies and technologies available in the 1970s and 1980s. Another lesson can be drawn from these early attempts at estimating tropical deforestation: deforestation rates calculated from highly uncertain forest area estimates are mere speculation. The experiences of the last two decades of the twentieth century will provide us with ample illustrations of this difficulty.

2

MONITORING TROPICAL RAINFOREST TRENDS

A Tortuous Exercise

THE EARLY 1990S marked a turning point in the international discourse on rainforests. By now the world had become aware of the threats to the future of these forests, the most important repositories of terrestrial biodiversity. The demise of the Tropical Forestry Action Plan, created to fight deforestation in 1985, was imminent (see box 2.1), and some countries considered the non–legally binding "Forest Principles" established at the 1992 United Nations Conference on Environment and Development in Rio de Janeiro to be a weak substitute for a legally binding intergovernmental agreement on forests.

After their experiences with the creation of the United Nations Framework Convention on Climate Change and the United Nations Convention on Biological Diversity, nongovernmental organizations were unsure about whether a forest convention would be helpful. But another important development at the Rio conference was a backlash against nongovernmental organizations that campaigned against deforestation in tropical countries. Those who attended the Rio conference will hardly forget the forceful blow of the Malaysian Prime Minister Mahathir against the "imperialist" denunciations from northern nongovernmental organizations and European forest ministers. As justified as northern criticism of rapid deforestation in places like Malaysia may have been, the rebuttal from the south was equally effective. Mahathir asserted that European countries were in no position to criticize the forest management of developing countries, having deforested their own countries to a much larger degree in earlier times. He successfully displaced the problem of tropical deforestation into the court of northern countries.

Disillusioned with the ineffectiveness of the intergovernmental process, some of the larger nongovernmental organizations began to look to independent certification as a solution to destructive logging and the abuse of indigenous rights. International conservation organizations—most prominently WWF—took the lead in establishing the Forest Stewardship Council (see box 4.1). Timber certification in the tropics, however, offered only a partial solution to deforestation and forest degradation. In the 1990s, most of the forest loss—about 70 percent—was through conversion to agricultural land. Most of this conversion was to permanent large-scale agriculture in Latin America and Asia, but some was to small-scale shifting cultivation, mainly in Africa.

BOX 2.1   **The Tropical Forestry Action Plan: A failed attempt?**

The Tropical Forestry Action Plan (TFAP) of 1985, sponsored by the World Bank, the World Resources Institute (WRI), the United Nations Development Programme (UNDP), and the FAO was to provide a strategy for the conservation of tropical forests. The plan built on the recommendations of an international task force convened by the WRI[1] and was intended to "increase action against deforestation from the narrow confines of the forest community to the wider arena of public policy." It was estimated that putting the plan into action would require an investment of about USD 8 billion of development aid money in the first five years, and involved about seventy countries in the effort. However, by 1990 critical reports had been published by the World Rainforest Movement,[2] and WWF International formally withdrew from the process in response to the criticism from some of their country offices. Even the WRI, one of the sponsors of the TFAP, expressed severe disappointment with the progress of the plan.

Notably, the critics included even the FAO itself, even though it had coordinated the initiative. The criticism, which was expressed in an independent review commissioned by the FAO, focused on the excessive emphasis on commercial forestry, a lack of analysis of the causes of deforestation, a failure to promote biodiversity conservation and sustainable forest management, and a lack of involvement of nongovernmental organization and local communities. The fiercest critics among the nongovernmental organizations went as far as to suggest that the plan would accelerate deforestation rather than curbing it. In 1991 the TFAP objectives were reformulated, with a name change to Tropical Forestry Action Programme, and the FAO's coordination role was distributed among various developing countries. But the restructuring process was so slow that the TFAP Forestry Advisors Group felt that the initial purpose of the TFAP had in the meantime been overtaken by the World Commission on Forests and Sustainable Development created after the 1992 Earth Summit in Rio de Janeiro.

In 1995 the advisors group proposed a framework concept for national forest programs in developing countries. The FAO itself recognized that the TFAP was clearly out of date by then, as many countries were trying to find their own way out of increasing deforestation rates.[3] Meanwhile, the Intergovernmental Panel on Forests (IPF) and later the Intergovernmental Forum on Forests (IFF) in 1997 were mandated to follow up on the "forest principles" decided at the Earth Summit. Based on the IPF/IFF outcome, the United Nations Forum on Forests was established in 2000 by the United Nations Economic and Social Council to promote the management, conservation, and sustainable development of forests.

The TFAP can hardly be considered to have been a success of any sort, not even by its initiators and sponsors. But it can be seen as the first step toward the United Nations taking a wider view of global forest issues and looking beyond the narrow concerns of the forestry sector. But it is open to debate whether the United Nations Forum on Forests and its predecessors (IPF/IFF) contributed to slowing deforestation, particularly in the tropics.

The FAO's Forest Resources Assessments

THE 1990S ALSO brought a more systematic effort to produce reliable data on forest area trends, especially for the tropical rainforest. The emergence of remote sensing techniques raised hopes of getting conclusive information on forest cover and deforestation rates. Without such information it would be difficult to understand the causes of deforestation and even more challenging to draw up policies against them. The FAO, which had been responsible for the forest resources within the United Nations system since 1948, was again the center of attention. Its Forest Resources Assessments, commonly known as the FRAS, which have been published since 1948 and are now published every five years, promised to establish the necessary data sets on the world's forest cover, including the tropical rainforests. FRA 1990 attempted to use forest cover data from different reporting dates to describe the FRA baseline year by back-calculating the data using independent variables such as human population density and growth, which were seen as major drivers for deforestation.[4] However, such simple models for estimating deforestation rates were shown to be flawed,[5] and the FAO subsequently discontinued such algorithms to project forest cover change.

THE FAO'S FOREST CLASSIFICATION SYSTEM: A MOVING TARGET

A NUMBER OF specialists have criticized the FRA data since the 1990s, most prominently Alan Grainger[6] from the University of Leeds in the United Kingdom. He pointed out that the FAO has changed its forest classification system in successive FRAS. In FRA 1980, "broadleaved forests" were classified as "closed," "open," and "bamboo" forest, separate from "forest plantations." In FRA 1990, "closed" and "open" forest were combined into "natural forest," and ten years later natural and planted forest (including rubber plantations) were merged into "total forest," before that class was again subdivided into "primary," "modified natural," and three different classes of planted forest in FRA 2005. Methods of quantifying forest cover change in FRA 2005 were not consistent between countries, and the definition of "forest" was based on land use rather than actual forest cover.

Grainger also scrutinized the FAO's habit of only projecting *deforestation* trends, when in fact certain countries may have undergone a *reforestation* trend. Perhaps even more problematic were the backward projections of forest cover from the latest FRA assessments, which produced different (higher) forest cover estimates for the previous reference years.

The FAO has often been criticized for its dependence on member states to provide forest data, even after remote sensing techniques had become available. This is certainly a justified criticism. We have to recognize that it is extremely difficult to estimate the areas of different forest types in some seventy tropical countries, which use a variety of classification systems. The FAO itself, in FRA 2000, stated that their data could not be directly compared to those from FRAS 1980 and 1990 because of the shortcomings described above.[7] Other inconsistencies in FRAS were the result of a recommendation by the Intergovernmental Panel on Forests to apply the low and inadequate 10 percent minimum tree cover as the definition of a forest in all countries of all climatic zones, including the tropical rainforest countries.

FRA 2010: COMPREHENSIVE BUT NOT VERY DETAILED

IT WAS A step forward when the latest Forest Resources Assessment, FRA 2010,[8] used essentially the same forest class definitions as the previous assessment (see box 2.2). FRA 2010, according to the FAO, is the most comprehensive assessment to date, based on 233 country reports from 1990, 2000, 2005, and 2010. However, taking into account that the FAO uses a minimum tree cover percentage of 10 percent and does not distinguish between humid and dry forest biomes, the area estimates and deforestation rates of the FRA 2010 are of limited value for monitoring tropical rainforests.

BOX 2.2   ## Terms and definitions used in the FAO's FRA 2010

The FAO classifies land as forest land and nonforest land, where land can be classified as forest land even if it is temporarily not covered by forest. Consequently, the FAO also distinguishes between "reforestation" (of forest land) and "afforestation" (of nonforest land). Because the FAO definition of forest land ignores the actual vegetation cover, the FAO data sometimes do not match forest cover data from remote sensing. Most authors, therefore, classify land by its actual vegetation cover instead of using the FAO's definition of "forest land." This also means that the term "reforestation" is often used to refer to any kind of forest cover increase whether through planting of trees, natural succession (regrowth), or natural expansion of forest.

afforestation. Establishment of forest through planting or deliberate seeding on land that, until then, was not classified as forest.

deforestation. Conversion of forest to another land use, such as agriculture, pasture, water reservoirs, or urbanization, or the long-term reduction of tree canopy cover below the 10 percent threshold.

enrichment planting. The planting of tree species in a natural forest with the objective of creating a high forest dominated by desirable tree species.

forest. An area of more than half a hectare covered with trees higher than five meters that produce a canopy cover of more than 10 percent, or trees able to reach this size and coverage. Both natural and planted forests, including rubber tree and pulp and paper plantations, count as forests.

forest degradation. Changes of forest cover through logging, fire, windfall, or other causes that negatively affect the structure or function of forests and lower its capacity to supply benefits, such as wood, biodiversity, and other ecosystem services.

forest fragmentation. The conversion of formerly continuous forest into patches of forest separated by nonforested land. This is usually a result of logging, subsistence farming, road building, or similar activities.

natural expansion of forest. Expansion of forest through natural succession on land that, until then, was under another land use (for example agriculture).

naturally regenerated forest. Forest predominantly composed of trees established through natural regeneration (regrowth) on land classified as forest.

planted forest (formerly called forest plantations). Forest predominantly composed of trees established through planting or deliberate seeding of indigenous or introduced species.

primary forest. The FAO characterizes primary forest as "naturally regenerated forest of native species, where there are no clearly visible indications of human activities and the ecological processes are not significantly disturbed." See also "secondary forest."

reforestation. Reestablishment of forest through planting or deliberate seeding on land classified as forest. Many authors use this term to describe any forest cover increase, whether human-induced or natural.

secondary forest. There is considerable confusion about the use of the term "secondary forest." The following working definition has been proposed: "Secondary forests are forests regenerating largely through natural processes after significant human and/or natural disturbance of the original forest vegetation ... and displaying a major difference in forest structure and/or canopy species composition with respect to nearby primary forest on similar sites."[9] Under this definition, forests subject to low-intensity selective logging or small-scale extractive activities, for example for nontimber forest products, are not considered to be secondary forests. See also "primary forest."

FOREST AREA: A GOOD INDICATOR OF FOREST HEALTH?

THE FAO STRESSES the point in FRA 2010 that forest area has often been overemphasized as an indicator of forest development and health, particularly in public debate. The FAO also states that in assessing land use dynamics, gains from afforestation and natural expansion of forests should be considered along with forest loss, and that the role of forests for carbon storage and growing timber stock should receive more attention. Although there is certainly some truth behind these assertions, they could also be considered to be evasive. As long as there are no reliable indicators of forest health, biodiversity, and ecosystem services, and no standardized measures for carbon storage, forest area remains the most important, if not the only, quantifiable indicator.

Ironically, although the FAO is concerned about the degradation and fragmentation that affect forest health and biodiversity in very large areas of the tropics, the FAO's own classification system fails to provide any meaningful information about forest health and devel-opment because it uses a minimum tree cover of only 10 percent to define forest area. Forest health throughout the tropical rainforests will therefore have to be assessed from data sources other than FRA 2010—data that use a considerably higher minimum tree cover and that use imagery of actual forest cover and not theoretical "forest land" that may not have any tree cover.

Emerging remote sensing technology

WHEN THE FIRST Landsat imagery was made available by NASA in 1972–73, it fueled hopes that the FAO tropical forest cover monitoring program would be able to produce accurate data within a few years. Initiatives were launched by the FAO, NASA, the International Union for Conservation of Nature (IUCN), the Woods Hole Research Center, and the European Commission Joint Research

Centre under their TREES project. All these remote sensing survey initiatives aimed to establish a reliable baseline of the tropical forest resources, but the lack of standardized criteria and definitions made the data from these survey initiatives difficult to compare.

THE TREES PROJECT: SOPHISTICATED METHODOLOGY TO ASSESS DEFORESTATION

PERHAPS THE MOST promising of these attempts was the large-scale TREES (Tropical Ecosystems Environment Observation by Satellite) project by the European Commission Joint Research Centre. Its goal was to provide accurate information on tropical forest ecosystems and to analyze deforestation trends.[10] One of the main motivations for the project was the uncertainty about deforestation rates in the tropics: the Intergovernmental Panel on Climate Change estimated that error rates could be as much as 50 percent. More accurate values were needed to assess carbon emissions from deforestation. TREES developed a new technique for measuring global tropical forest cover, based on satellite imagery with a resolution of one square kilometer. It was produced by the advanced very high resolution radiometer (AVHRR) from the National Oceanographic and Atmospheric Administration (NOAA) and complemented with selected medium-resolution Landsat thematic mapper data for correction and validation (for an overview of satellite sensors, see appendix 2). The project also produced the first detailed remote sensing maps at a scale of 1 to 5 million for Central Africa and Madagascar, South America, and Southeast Asia.

The TREES-2 project (1996–2000) subsequently analyzed deforestation and forest degradation using the latest remote sensing information for the humid tropical forest ecosystems.[11] First, the previous area estimates were slightly revised and supplemented with confidence limits. A comparison with the estimates extracted from the country tables of FRA 1990 and with estimates from the

IUCN regional atlases[12] showed close similarity in the global figures. But the global totals also blurred some important differences in the regional estimates (see table 2.1). Then a change assessment was carried out using medium spatial resolution at each of the hundred hot spot areas of deforestation, nearest to the target dates of 1990 and 1997. To identify the change that had happened between these target years, the project applied two criteria to define forest cover: canopy density (crown cover) and forest proportion within a mapping unit. The resolution of the satellite imagery was high enough to allow a rational separation of closed forest from degraded and fragmented forest (see appendix 3).

TABLE 2.1. Tropical rainforest cover estimates for 1990, in millions of hectares

The "evergreen and seasonal humid tropical forest" category covered by TREES[13] corresponds closely with the FAO's "closed broadleaved forests" of FRA 1990 as taken from the country tables, and the IUCN category of "tropical closed forests."[14]

Figures for Latin America exclude Mexico and Atlantic forests in Brazil. Figures for Africa include West and Central Africa and Madagascar. Southeast Asia includes humid tropical forests in India and seasonal forests of continental Southeast Asia (Myanmar and Thailand).

	TREES	TREES-2	IUCN CLOSED FOREST	FRA 1990 CLOSED BROADLEAF FOREST
Africa	207	198±13	199	218
Latin America	671	669±57	693	652
South and Southeast Asia	281	283±31	271	302
TOTAL	1,158	1,150±54	1,163	1,172

The TREES project's change assessment for 1990–1997 set a new standard for forest cover change estimations. It established what is

likely to remain the most reliable data for the 1990s: a global tropical rainforest cover of 1,150 ± 54 million hectares in 1990, and an
annual gross change of 5.8 ± 1.4 million hectares in this period,
corresponding to an annual loss of 0.52 percent. Taking reforestation of 1.0 ± 0.32 million hectares into account, the annual net loss
amounted to 4.9 ± 1.3 million hectares, or 0.43 percent annual net
cover loss (see appendix 3). This net deforestation rate calculated by
the TREES project was about one-fourth lower than the global net
deforestation calculated from the FRA 2000 data.

THE ABSENCE OF STANDARDS

THE TREES PROJECT showed clearly that the discrepancies in the
FAO statistics were caused by inconsistent definitions. The first FAO
remote sensing survey was less useful because it used such a low
canopy cover threshold to define closed forest areas: the minimum
of 10 percent crown cover is not a good definition of forest in the
case of tropical rainforests. The absence of a specific class for fragmented and degraded forests also led to great uncertainty in the
interpretation of satellite imagery. Remote sensing is not an ultimate answer to the complex question of forest cover analysis. Even
the currently used remote sensing technology makes it difficult to
distinguish between primary forests, older secondary forests, and
tree plantations.[15]

Inaccuracy in data and a lack of common definitions translate
into scientific uncertainty about biodiversity, carbon stocks, and
the validity of such newer mechanisms as the United Nations
program Reducing Emissions from Deforestation and Forest Degradation (REDD). Before these policies can be applied, discrepancies
in the forest classification systems for regional and national forest
inventories have to be straightened out, recognizing that perfect
accuracy is an illusion.[16] Multilateral, state, and academic agencies
have difficulty finding a commonly accepted way to deliver reliable,

comparable remote sensing forest data, and in the end this gets in the way of drawing meaningful policy conclusions.[17] The importance of the tropical rainforests would justify creating an institution equivalent to the Intergovernmental Panel on Climate Change to handle forest cover monitoring and standard setting. It has become evident that the FAO cannot fulfill this role.

COMPUTING GLOBAL DEFORESTATION RATES

GLOBAL DEFORESTATION RATES computed using area estimates from different periods are very unreliable. Forest cover maps based on satellite imagery from different sensors in different periods likewise cannot be used to compute global forest cover change. The inaccuracy in forest area estimates is simply too large to allow quantified comparisons over time periods of five or ten years: the margin of error in the estimates is bigger than the actual changes. The famous analogy of football fields lost per minute on the global scale is therefore as good as any, considering the different sizes of football fields.

The amount of error in forest cover measurements has even led some authors to question whether the global rainforest area was actually decreasing. Studies of natural regrowth and forest expansion from a number of tropical countries further increased these doubts. The only way to assess forest cover change is to do a pixel-by-pixel comparison of maps made with the same satellite imaging technology and interpretation.

New horizons in remote sensing technologies

REMOTE SENSING TO monitor vegetation cover has become a highly sophisticated science in itself, with dozens of different satellite sensors in use and new ones being launched increasingly often. The need for standards for forest cover monitoring has been partly filled by remote sensing scientists. Since the beginning of this century this has greatly improved understanding of the dynamics of deforestation at the national, regional, and global levels. Remote sensing technology, however, has become difficult to understand for the nonspecialist, who is often at a loss when trying to compare data from different origins recorded with different sensors.

Today there are over sixty satellites carrying imaging systems from more than twenty-five countries. Many of these imaging systems could provide suitable imagery for forest monitoring (see appendix 2) if policies for access, processing, and distribution existed.[18] Remote sensing imagery can be very expensive, which is one of the reasons why most forest cover mapping has thus far been based on the freely available moderate resolution imaging spectro-radiometer (MODIS) imagery. Over a period of ten years, dozens if not hundreds of scientific papers have been published in specialized journals dealing with remote sensing of tropical forests. With the relative proximity of specialized United States universities and other institutions, a majority of these papers dealt with deforestation in the Amazon Basin. Some useful overviews of remote sensing technologies used in forest monitoring have been published,[19] but these become outdated quickly when new sensors and interpretation techniques are launched. New sensors improve accuracy and produce more detailed imagery, but these new data come with the disadvantage that they cannot be compared with earlier data.

THE PROBLEM OF CLOUD COVER

A NOTORIOUS PROBLEM with optical sensors is the persistent cloud cover over certain tropical rainforest areas. This has hampered the monitoring of forest cover change in some Indonesian regions with high rates of deforestation, such as Sumatra and Kalimantan. Because cloud-free observation opportunities do not happen as often, it is easy to miss transient forest cover change. When primary forest is cleared and replaced by tree plantations, this change in forest cover type is not easy to distinguish with less frequent pictures. Plantations of oil palms, rubber, and other crops have been promoted in Indonesia at the expense of natural forests, and such changes can easily be missed. A number of leading remote sensing specialists have tried to get around this problem, while still relying on the freely available Earth Resources Observation and Science (EROS) Landsat archive of the US Geological Survey. They analyzed the continuous change per pixel, integrating low-resolution MODIS data with medium-resolution Landsat 7 data.[20] For persistently cloudy tropical forest areas, this method produced better results than comparing a variety of images from specific years.

FOREST COVER MONITORING HELPS SLOW DEFORESTATION IN BRAZIL

REMOTE SENSING SPECIALISTS now commonly use low-resolution imagery to identify deforestation "hot spots" combined with medium-resolution imagery from Landsat to assess the precise extent of forest cover change in these areas. A combination of low- and medium-resolution analysis is particularly useful in areas with diffuse patterns of forest cover change, and it is commonly found in slash-and-burn farming areas in Central Africa.[21]

The most successful use of forest cover monitoring was in Brazil (see box 2.3). In 1988 the Brazilian government initiated the Amazon Deforestation Monitoring Project (PRODES), the world's most extensive national forest monitoring program. PRODES is essentially based on Landsat imagery, combined with DETER, a real-time detection of deforestation hot spots, based on low-resolution MODIS data.[22] It was designed to implement the Brazilian government's plan for preventing deforestation in the Brazilian Amazon region. PRODES has undoubtedly contributed to the impressive results in slowing deforestation in the Brazilian Amazon since 2004.

BOX 2.3   **A new deforestation analysis for the Amazon biome 2004–2012**

CLÁUDIO C. MARETTI

Most deforestation data and analyses available do not consider the Amazon as an ecological unit and are not based on comprehensive data from all Amazon countries. Although the official Brazilian INPE-PRODES system offers excellent data, they apply to the legal, political-administrative definition of the Amazon, which includes parts of other biomes. The other Amazon countries are developing their own monitoring systems, but in most cases they are not yet consistently defined and applied. This new analysis was inspired by Coca-Castro et al.[23] and is based on data from the following sources:

- The WWF Living Amazon Initiative[24] has developed a definition of what the Amazon is that considers mostly the biogeographic limits. Besides the dominant moist broadleaf forests, the Amazon biome includes other vegetation types such as floodplains, swamps, bamboo, and palm stands, but not dry or high mountain woodlands or savanna areas.
- Terra-i[25] provides data on deforestation. The natural habitat conversion

Gross deforestation in the Amazon per country, in hectares, 2004–2012

	2004	2005	2006	2007	2008
Bolivia	29,530	72,684	59,384	46,873	75,972
Brazil	1,083,806	1,348,459	1,598,245	959,938	1,863,385
Colombia	36,973	23,133	30,259	60,129	65,416
Ecuador	10,100	4,036	1,854	3,642	4,401
Fr. Guiana	968	1,288	2,263	1,981	2,887
Guyana	892	1,730	2,670	3,579	6,265
Peru	38,023	55,523	34,107	35,152	62,781
Suriname	2,650	2,394	2,644	2,431	5,500
Venezuela	12,747	9,898	12,237	29,328	20,285
TOTAL	1,215,689	1,519,145	1,743,663	1,143,052	2,106,892

data from Terra-i are an easily accessible source for the whole Amazon biome. Data per country were organized according to the ecoregions of the Amazon biome for each country.

The results of this new analysis suggest an accumulated deforestation in the entire Amazon biome from 2004 to 2012, to amount to some thirteen million hectares (see table below). In absolute terms, Brazil suffered a higher loss due to deforestation than all the other countries together, accounting for 82.4 percent of deforestation. The Andean Amazon countries accounted for 16.6 percent, and the Guianas accounted for a little more than 1 percent of the total deforestation.

However, in relative terms, the deforestation lead of Brazil could soon be a thing of the past if Brazil continues to curb deforestation and the Andean Amazon countries do not correct their current tendencies. Bolivia in particular has shown alarming deforestation rates, particularly in 2010 and 2011,[26] and Peru also shows increasing deforestation. Thus, even in the Andean Amazon countries and the Guianas, with their lower historical deforestation record, the Amazon forests are now coming under increasing pressure. In the past two decades, these countries

2009	2010	2011	2012	2004–12
28,421	395,409	196,882	32,047	**937,202**
574,071	1,255,545	1,319,442	760,596	**10,763,488**
45,410	56,943	27,111	10,018	**355,391**
4,046	8,817	11,930	5,436	**54,262**
1,163	3,131	2,869	3,732	**20,281**
5,313	14,364	8,729	19,112	**62,653**
69,259	70,179	94,633	109,317	**568,974**
5,488	8,375	10,115	18,305	**57,896**
21,351	84,369	22,906	37,103	**250,224**
754,520	**1,897,132**	**1,694,618**	**995,665**	**13,070,371**

have increased their exploitation of their part of the Amazon with oil and gas, mining, and more commercial agriculture and cattle ranching. A demographic movement from the Andes to the lowlands seems to be happening, driven by economic factors. Peru has experienced important economic growth, but at the cost of opening virtually the whole country for mining and oil and gas exploitation. Bolivia seems to be combining a movement of people from the Andes into the forested lowlands with the adoption of large plantations and cattle ranching, following the Brazilian example of the past decades.

The Andean Amazon countries have also adopted a number of measures to reduce deforestation. Peru committed to reduce emissions from land use change to zero by 2021. Bolivia develops its own climate mitigation and adaptation mechanism and the "Madre Tierra" legal framework (a law introduced by the president of Bolivia, Evo Morales, that considers Mother Earth as sacred and aims at development in harmony with nature). Colombia has a commitment to zero net deforestation by 2020. Ecuador enhanced its budget for the national system of protected areas. And the Amazon Cooperation Treaty Organization coordinates deforestation monitoring. But despite these first attempts, none of the Andean Amazon countries seem to be consistently executing policies and programs to reduce deforestation in the Amazon yet. Even Brazil's impressive efforts in the last decade to curb deforestation have not yet brought this country to a low enough deforestation level.

Therefore, we still need to develop new models of low deforestation and low carbon emission for all the Amazon countries. With the exception of the Amazon Region Protected Areas Program in Brazil, efforts by the international community have been largely confined to pilot projects and nonmarket approaches and were very limited in scope, scale, and reach. International agencies and the investment sector in industrialized countries should collaborate with the Amazon countries on policies and programs to reduce deforestation and enhance conservation, and scientific research should investigate and test more consistent sustainable development options for the Amazon countries.

CRUCIAL NEW REMOTE SENSING TECHNOLOGY

IMPLEMENTING THE REDD program will require better techniques for measuring biomass and carbon stocks. It will require active satellite technology such as synthetic aperture radar (SAR) and light detection and ranging laser (lidar) (see appendix 2). Assessments will require frequently updated "wall-to-wall" coverage in sufficiently high resolution, as well as ground truthing (verifying the satellite interpretation of the vegetation type on the ground, to make sure that, for example, closed forest is not confused with a tree plantation). The Copernicus program of the European Space Agency specifically aims to help bridge this gap with its Sentinel satellite series. A new SAR satellite, DESDynI, is also expected to be launched by the United States in 2019. This satellite will integrate SAR and lidar technology and should produce accurate measurements of ecosystem structure (forest density, biomass, whether a forest is natural or planted) and forest height.[27]

A comprehensive overview and analysis of remote sensing methods for global forest monitoring has been compiled in a recent book, *Global Forest Monitoring from Earth Observation*, by two of the leading specialists in this field: Frédéric Achard, senior scientist at the Joint Research Centre of the EU Commission, and Matthew C. Hansen, professor in the geographical sciences department at the University of Maryland.[28] But progress does not stop there. Shortly after the publication of this overview, at the end of 2013, Hansen et al. published an interactive global forest cover change map using Google Earth Engine.[29] It has the potential to revolutionize forest cover monitoring!

The "Traveler's Tree" (*Ravenala madagascariensis*) is an endemic species of Madagascar, today common as an ornamental plant in many tropical countries.

3

GLOBAL AND REGIONAL
DEFORESTATION PATTERNS

MUCH HAS BEEN said and written about the destruction of tropical rainforests over the past forty years, and a lot of energy and resources have gone into establishing reliable area estimates and deforestation rates. Mistakes and wrong assumptions have stalled progress, and the most pervasive of these has been the belief that it would be easy to define what a tropical rainforest is, that its area could be measured to the hectare, and that remote sensing would bring all the solutions.

Around the turn of the millennium a decisive shift took place in the tropical rainforest debate. Concern over the loss of tropical forests and their rich biodiversity was partly replaced by concern about the effect of deforestation on the global climate. Climate change was still being denied or ignored in business circles and in some of the most powerful governments, but the Kyoto Protocol of December 1997 reflected a growing worldwide concern. The Kyoto Protocol created a requirement for an inventory of greenhouse gas emissions from human-induced land use, land use change, and forestry activities, known under the acronym LULUCF.

The role of deforestation in climate change had not been taken into account until it re-emerged in 2005 at the United Nations Framework Convention on Climate Change Conference of the Parties (COP II) in Montreal. The Coalition for Rainforest Nations, led by Papua New Guinea and Costa Rica, came up with the REDD (Reducing Emissions from Deforestation and Forest Degradation) proposal for developing countries. The resulting Bali Action Plan of December 2007 proposed incentive payments to reduce emissions from deforestation and forest degradation. A later version of REDD, REDD+, took into account contributions from conservation, sustainable management of forests, and enhancement of forest carbon stocks.

The focus of public concern and discussion about rainforest destruction had been partly displaced from concern about the impact of deforestation on nature and people to its effect on climate, which many people saw as being more important than loss of biodiversity. Many people still found biodiversity a difficult topic to understand, let alone quantify. Increasing concerns over the greenhouse gas emissions from deforestation, which the fourth IPCC assessment report of 2007 estimated at 17.3 percent of all human-induced emissions, meant that tropical forest cover change was now primarily seen in this light. This in turn created a requirement for more precise forest cover estimates and ultimately remote sensing technology that could assess biomass and carbon stocks.

Global tropical rainforest cover change from 2000 to 2010

THE PIONEERING WORK on satellite imagery during the 1990s by the TREES project and a few others benefited from the open access low-resolution AVHRR and MODIS data, as well as the medium-resolution data from Landsat (see appendix 2). Global forest cover assessment science would not have made the progress that it did over the preceding twenty years without the free and open access policy of NASA and the US Geological Survey.

When the United States government subsequently adopted an open access policy for its Landsat archives in 2008, it allowed the scientific and environmental communities around the world to base their programs on accurate forest monitoring data and to assess deforestation rates more accurately, potentially using information going back to the 1970s.[1] An important benefit of free access to historical Landsat data was that forest cover change could be assessed using images made with the same methods, which increased the accuracy dramatically. The time of assessing forest cover change using forest distribution maps made with different methods and based on different forest classification systems had now ended.

SAMPLING DEFORESTATION HOT SPOTS

IN THE FIRST decade of the new millennium it was not feasible to analyze the global rainforest cover change with full-scale medium-resolution coverage. Such a vast undertaking would have meant monitoring billions of Landsat pixels. Instead, researchers used sampling techniques similar to those developed by the TREES project. To assess forest cover change from 2000 to 2005, Hansen et al.[2] used a sampling approach based on low-resolution satellite data from MODIS. The advantage of low-resolution data from MODIS, or the vegetation sensor of the SPOT satellites (see appendix 2), is that these satellites produce new images almost every day, which makes cloud cover much less of a problem. A further advantage of low-resolution imagery is that it can be used to detect deforestation hot spots.

To capture small-scale forest clearing within a deforestation hot spot, Hansen et al. used medium-resolution Landsat data. They analyzed a stratified random sample of 183 squares, 18.5 kilometers on each side, for forest clearance with Landsat imagery for the years 2000 and 2005 respectively. The deforestation hot spots identified for the 2000–2005 period (see plate 2) showed that 47.8 percent of deforestation in tropical rainforests happened in Brazil, predominantly in the states of Acre, Rondônia, Mato Grosso, and Pará.

Indonesia showed the second-highest degree of forest clearing: 12.8 percent. The massive deforestation in both these countries was mostly caused by forest conversion for agro-industrial purposes. Humid tropical Africa, on the other hand, accounted for only 5.4 percent of deforestation, despite widespread selective logging. The surprisingly low percentage of deforestation in Africa reflected the virtual absence of agro-industrial plantations during this period. Another notable feature of the map produced by Hansen's group was that the forest area that showed low or no deforestation in Southeast Asia was much reduced, showing that the primary forest areas in that region were rapidly shrinking and fragmenting.

There was still controversy about sampling techniques for forest areas, even with consistent methodology. Some authors questioned whether sampling should ever be done, and called for wall-to-wall, comprehensive coverage instead.[3] Because of the cost and other practical considerations, wall-to-wall coverage at sufficient resolution for the entire tropical rainforest domain seemed unthinkable—at least until very recently.

THE GLOBAL FOREST MAP AT 30-METER RESOLUTION

AT THE END of 2013, *Science* published a paper by Matthew C. Hansen from the University of Maryland with a number of coauthors.[4] It included a link to a global forest map that is likely to revolutionize remote sensing of forest cover and its change over time: http://earthenginepartners.appspot.com/science-2013-global-forest.

This map was created using Google Earth Engine and is based on 650,000 Landsat 7 satellite images consisting of 143 billion pixels. It charts the world's tree cover and its loss and gain for the period of 2000–2012 at a resolution of thirty meters and will be updated annually. For the first time, a dynamic global wall-to-wall forest cover assessment is available. The map's tracking of historical data allows it to detect the replacement of natural forest with tree plantations, something which in the past has led to distortions of forest statistics, notably in Indonesia.

The high-resolution global forest cover change map brought the sobering proof that deforestation is not only a tropical phenomenon. Globally 229 million hectares were lost and 80 million hectares were gained through reforestation during this period. For the first time a major loss of forest cover was also documented for temperate and boreal countries: Russia in the first place, followed by the United States and Canada. Forest loss was largely driven by intensive logging, fires, and tree disease.

The global forest map analysis differentiated between subtropical and tropical climatic areas. For the tropical domain three biomes were distinguished: tropical rainforests, tropical moist deciduous forests, and tropical dry forests. A compilation of tropical rainforest gain and loss (see table 3.1) allows a comparison with the global figure for tropical rainforests for 1990–1997 by Achard et al.[5] The areas surveyed may not match precisely, since the recent high-resolution map picked up smaller forest areas in some places, such as on some Pacific islands. However, the trends of forest loss are clear. The percentage of tropical rainforest lost in the 1990s decreased from 0.52 percent per year to 0.43 percent in the first twelve years of this century (gross loss). In other words, in the 2000–2012 period we still lost about 4.9 million hectares of rainforest (with more than 50 percent tree cover)—about 1.2 times the size of Switzerland—per year.

TABLE 3.1. Tropical rainforest areas 2000–2012

Tropical rainforest (TRF) areas with more than 50 percent tree cover in 2000 and forest areas lost and gained between 2000 and 2012, in millions of hectares. Extracted from supplementary materials in Hansen et al. (2013).[6]

	TRF TREE COVER IN 2000	TRF TREE COVER LOSS BY 2012		TRF TREE COVER GAIN BY 2012	
	Mha	Mha	12-year loss in %	Mha	12-year gain in %
South America	573.9	24.7	4.3	3.3	0.6
Central America and Caribbean	29.0	2.2	7.6	0.3	0.9
South and Southeast Asia	193.7	22.2	11.5	10.4	5.4
Pacific and Australia	71.9	1.0	1.4	0.3	0.5
West and Central Africa	268.4	8.4	3.1	2.5	0.9
TOTAL WORLD	**1,136.9**	**58.5**	**5.1**	**16.8**	**1.5**

Rainforest gains through reforestation, on the other hand, increased from 0.08 percent per year in the 1990s to 0.12 percent (equivalent to 1.4 million hectares per year) since the beginning of this century. The two inverse tendencies amounted to a net annual loss of 0.43 percent in the 1990s and 0.31 percent net loss since then. These net figures, though, are approximations because some areas may have lost forest and subsequently gained regrowth. The general tendency, nevertheless, is a slowing of deforestation rates, although as appendix 4 demonstrates, the global figures mask important regional differences in forest cover change.

Diverging regional deforestation trends

REGIONAL FOREST COVER trends are dominated by the deforestation trajectories of the largest rainforest countries in each rainforest region—Brazil in Latin America, the Democratic Republic of the Congo in Africa, and Indonesia in Southeast Asia. In the FRA 2010 report published by the FAO, estimates of deforestation rates were based on backward-corrected forest area estimates for the years 1990, 2000, 2005, and 2010, which took reforestation (natural and afforested) into account.[7] Since FRA 2010 made no distinction between forest types, the regional as well as country compilations include both humid tropical and drier forest types. They also include open forests because of the low minimum tree cover percentage used by the FAO to define the forest class.

The area estimates and deforestation rates from FRA 2010 should therefore not be used as indications of tropical rainforest cover change. They are referred to here only because FRA 2010 data are still seen as the official tool for policy making. For this analysis, only the FRA 2010 data of countries that fall entirely, or at least predominantly, within the humid tropical forest biome are used (see appendix 4). The correspondence between the FRA 2010 net deforestation rates from this selection of rainforest countries and the deforestation rates of the high-resolution global forest cover change map is relatively close for Latin America–Caribbean and Africa (see figure 3.1). On the other hand, we can see a very substantial deviation in deforestation rates in the case of the Asia–Pacific region.

FIG. 3.1. Annual deforestation rates 1990–2012 in percent

Global and regional gross deforestation rates from Achard et al. for 1990–1997,[8] and Hansen et al. for 2000–2012,[9] based on multi-temporal satellite imagery: thick lines. Regional net deforestation rates (i.e., gross deforestation plus regrowth and afforestation) extracted from country tables of FRA 2010:[10] thin lines. For the computation of these rates from FRA 2010 country tables, see appendix 4. Although the FRA 2010 rates are not directly comparable with the gross deforestation rates from Achard and Hansen, for reasons explained in the text, they still exhibit questionably low deforestation rates for the Asia–Pacific Region since the year 2000.

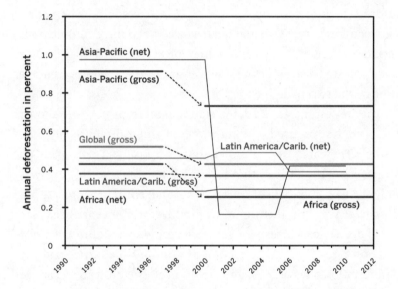

ASIA–PACIFIC: INDONESIAN STATISTICS HIDE THE TRUTH

THE STRIKING DIFFERENCE between the FAO deforestation statistics and those from independent multi-temporal satellite imagery confirms many people's subjective impression that the FAO data substantially underestimated the deforestation trajectory in the region.

The difference is mostly due to the deforestation statistics from the largest rainforest country in the region—Indonesia. According to FRA 2010 statistics, Indonesia appears to have decreased

Rainforest area in western Ghana cleared with slash-and-burn technique by immigrant farmers. (Photo by C. Martin.)

its massive forest loss of the 1990s of 1.75 percent per year to only 0.31 percent per year in the 2000–2005 period and 0.71 percent per year in the 2005–2010 period. These rates very likely misrepresent the actual situation; it has to be recognized that changes in the rate of forest loss in Indonesia have not been analyzed officially. An effective analysis would probably require a monitoring system comparable to Brazil's National Institute for Space Research (INPE), which has not been developed.

The annualized statistics of the new global forest cover change map[11] show a completely different picture for Indonesia. The country in fact experienced the largest increase of forest loss of all countries in the world since 2000 (see figure 3.2). This disappointing news corroborates the results of a detailed previous analysis by independent remote sensing specialists who used multi-year pixel change paths to detect gross forest cover loss. It revealed that Sumatra and Kalimantan together lost 5.39 million hectares or 1.15 percent annually of forest cover between 2000 and 2008—more than the area of Costa Rica.[12] Almost half this loss happened in just two provinces: Riau in Sumatra (see plate 7) and Central Kalimantan.[13] A very recent analysis also showed that the largest part of the Indonesian forest loss from 2000–2012 happened in primary forest areas.[14]

FIG. 3.2. Annual forest loss for Indonesia and Brazil 2000–2012 in millions of hectares

The two countries show an inverse trend of total forest loss (including all tree cover classes). The areas lost in Brazil are from all forest areas, not only the Brazilian Amazon, and are therefore higher, but the decrease of deforestation is largely due to the decline of deforestation rates in the Brazilian Amazon. Adapted from Hansen et al.[15]

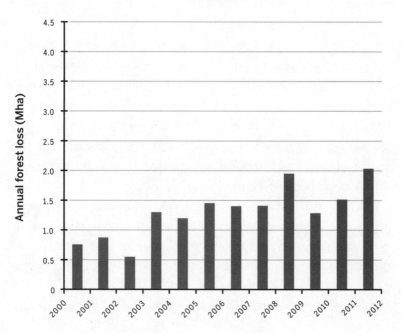

INDONESIA

BRAZIL

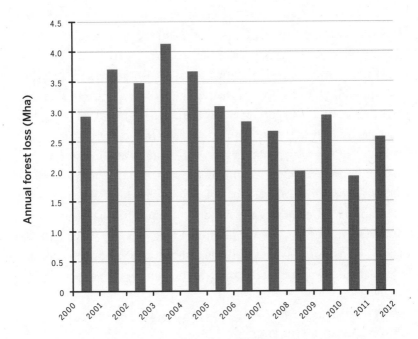

FRA 2010, on the other hand, listed a loss for the whole country of Indonesia of only 4.97 million hectares for 2000 to 2010. Such a substantial difference lies beyond a reasonable margin of error and needs to be explained if the suspicion of politically motivated misclassification is not to be raised. Unfortunately the FAO does not provide any satisfactory explanations. Part of the deviation may be due to the FRA 2010 figures being based on total forest area estimates, which means that they take regrowth and newly planted forests into account. But a major part of this difference is probably because of the shortcomings of the FAO classification system referred to in the previous chapter.

In reaction to the obvious increase of deforestation, the Indonesian government signed an agreement with Norway in 2011 to reduce carbon emissions. In return for a contribution of up to

USD 1 billion, the Indonesian government committed to a two-year moratorium on licensing agricultural plantations and logging concessions in intact natural forest and peatland. The results of this agreement have yet to be evaluated.[16]

Other Southeast Asian countries have also contributed to the continued high forest loss of the region. Malaysia showed an even higher percentage of forest loss in the 2000–2012 period (1.34 percent per year in the more than 50 percent tree cover category). In absolute terms, however, this loss was only about a quarter of the loss in Indonesia. Other countries in the region had a dampening effect on the net forest loss of the region: the Philippines and Vietnam had already lost the majority of their forests in previous decades, but had regained some forest area since 1990. Chapter 5, "Forest Transitions," will discuss these recent developments in some of the smaller tropical forest countries further.

SOUTH AND CENTRAL AMERICA AND THE CARIBBEAN: BRAZIL LEADS THE WAY

ACCORDING TO FRA 2010 as well as the high-resolution global forest cover change map, the annual deforestation rate in this region decreased slightly in the 2005–2010 period (see appendix 4). The deforestation rates were dominated by the trends in Brazil, where annual deforestation rates were slowing. In 1988 Brazil's National Institute for Space Research, INPE, under its monitoring program PRODES, estimated the total area deforested in the Brazilian Amazônia Legal region, which includes most of the country's tropical rainforests. It reported a deforested area of 28 million hectares out of an original area of about 409 million hectares, at an average annual rate of 2.1 million hectares between 1978 and 1988. This suggests that three-quarters of the total deforestation in the Amazônia Legal region happened between 1978 and 1988: over ten years, a forested area the size of Guyana disappeared.

An independent analysis of Landsat imagery for the entire Brazilian Amazon, however, arrived at a lower annual deforestation rate of 1.52 million hectares over the same decade.[17] Most of this difference could be explained by the fact that the INPE figures used a different limit to distinguish between closed forest and cerrado (wooded grasslands) in the states of Mato Grosso and Tocantins and a different estimation of secondary growth, which is difficult to distinguish in Landsat imagery.

Deforestation between 1978 and 1988 was concentrated in the "arc of deforestation" on the southern fringe of the Amazon Basin from the states of Maranhão through Tocantins, Mato Grosso, Rondônia, and Acre. It was triggered by the Brasília-Belém road (BR-010), followed by the BR-364 through the states of Mato Grosso, Rondônia, and Acre, as well as the PA-150 in Pará (see plate 2). Tax incentives for large-scale producers to convert forests to pastures were the main drivers of deforestation, but mining and selective logging also contributed.[18] In 1991 deforestation dropped to about half the average of the 1980s but increased again thereafter, peaking in 1995 and 2004 with close to three million hectares net forest loss. After 2004, deforestation rates in the Amazônia Legal region dropped as a result of the governmental action plan for the prevention and control of deforestation. Of all the countries in the world, Brazil showed the largest decrease in annual forest loss (see figure 3.2). Because of the size of the Brazilian Amazon, Brazil still showed the largest loss of rainforest areas in absolute terms, and the percentage of forest loss of 6.4 percent (in areas of more than 50 percent tree cover) over the 2000–2012 period was higher than the average South American rainforest loss of 4.3 percent. Some Central American countries, however, showed much higher rates, notably Guatemala (12.2 percent) and Nicaragua (11.2 percent).

After ten years of steady decline of deforestation rates in the Brazilian Amazon, 2013 marked a slight increase of 28 percent over the 2012 figure, but was still the second lowest since records started in 1988. Some environmentalists and media attributed this to the revised Brazilian Forest Code, which was passed in 2012 as a result of a farmers' lobby. But a commentary on Mongabay, one of the world's most popular environmental science and conservation news sites, considered this to be an unlikely hypothesis.[19] The commentators considered rising commodity prices and the absence of positive incentives for farmers to avoid land clearing to be more likely causes.

WEST AND CENTRAL AFRICA: THE WORST STILL TO COME?

FOREST STATISTICS FOR Central African countries have been notoriously uncertain. Estimations of the rainforest area of the largest country in the region, the Democratic Republic of the Congo, used to be approximate at best, and as a result the deforestation rates given in FRA 2010 are equally uncertain. They show a constant annual forest loss of about 0.45 percent between 1990 and 2010. Interestingly, if we only consider countries that fall mostly or entirely within the tropical rainforest zone, the annual deforestation rate decreases to about 0.30 percent (see appendix 4). This signals that dry forest zones in West and Central Africa are being deforested more quickly than the closed humid forest biomes. At least in Africa, it is evident that deforestation is not only a tropical rainforest phenomenon, but affects all forest areas across the continent. The high-resolution global forest cover change map has now confirmed a comparatively low rate of rainforest loss of 0.26 percent (in areas of more than 50 percent tree cover) per year for 2000–2012. If forest gains through reforestation were taken into account, the rate would be even lower.

In contrast to the other regions, the largest tropical rainforest country in Africa did not greatly influence the regional average. The Democratic Republic of the Congo showed practically the same annual deforestation rate of 0.27 percent. Some West African countries showed much higher rates of loss, particularly Côte d'Ivoire, which had already lost the majority of its forests in the past century.

A detailed change assessment for the Democratic Republic of the Congo between 2000 and 2010 was done using a wall-to-wall analysis of almost 9,000 Landsat ETM+ images to establish the baseline forest cover, as well as the gross forest cover loss between 2000 and 2010.[20] The 159.5 million hectares of total forest cover estimated was consistent with the estimate in FRA 2000. This figure, however, also included drier woodlands, leaving only 123 million hectares of tropical rainforest. Of this area, primary forest amounted to 105.3 million hectares. In the 2000–2010 period only 0.1 percent of primary forest was lost, whereas a high 1.2 percent of secondary forest, amounting to 17.5 million hectares, was lost annually.

The higher than average deforestation in secondary forests could mainly be attributed to subsistence agriculture, partly because logging roads provided easier access to forest areas (see plate 6). Charcoal production and mining also contributed to the deforestation. The wall-to-wall analysis of the Democratic Republic of the Congo forest cover also detected an increase in deforestation in 2005–2010. This increase was greatest near densely populated areas.

BOX 3.1

Forest conversion to agro-industrial plantations:
An upcoming problem in Central Africa

RALPH M. RIDDER

When chain saws came to Central Africa in the 1960s, they brought six major forest management problems with them.

1. **Illegal logging in and around approved forest concessions**

   Illegal logging in approved concessions takes place because of poor forest governance.[21] This problem can be solved with private legality audits by international auditing firms and by forest certification using Forest Stewardship Council or equivalent standards.

2. **Forest degradation in approved forest concessions**

   Many commercial tree species harvested during the first logging rotation do not fully recover within a twenty-five- to thirty-year logging cycle.[22] The consequence is a much poorer harvest volume in the second rotation, unless the market switches to lesser-known species.

3. **"Informal sector" logging by small and medium enterprises**

   Small and medium enterprises operate in a legal void because forest policy only covers operations by large enterprises. In many countries the total timber volume exploited by smaller enterprises is many times the volume harvested by the "formal" large concession holders.[23]

4. **Commercial bushmeat trade**

   Often organized by the government officials who are officially responsible for wildlife protection, the commercial bushmeat trade depletes wildlife across Central Africa. Hunters shoot wildlife with semiautomatic weapons to supply local and international markets with meat and Asian markets with ivory.[24]

5. **Low market demand for legal and certified tropical timber**
   Many consumers avoid buying any timber from natural forests in an effort to avoid buying illegal timber, but this unintentionally damages responsible tropical timber production as well.[25]

6. **Insufficient creation of timber plantations**
   Timber plantations can produce cheap fuelwood and construction timber, which takes illegal logging pressure off natural tropical forests.[26]

These pressures have been challenging to deal with, but today we are facing a much bigger problem. Over the last few years, severe pressure has been building from large-scale agro-industrial corporations competing for forest land to produce palm oil, cocoa, pineapple, coffee, and other cash crops. These companies could convert at least 30 percent of the current tropical rainforests in Central Africa to agricultural plantations in the next fifteen to twenty years.

Multinationals are discovering Central Africa, a large arable land area with, thus far, low population densities and negligible potential for conflict with current land users. The only obstacle seems to be the tropical forest cover, which has to be converted by clear-cutting, and the risk that end consumers will resist this. This is a tricky issue, and the multinationals want to avoid mistakes like those committed by donor agencies and industrial agribusinesses in Côte d'Ivoire, where the natural tropical forest cover was reduced from an original fourteen million hectares in the 1960s to about one million hectares, essentially limited to protected areas.[27] The massive deforestation that made Côte d'Ivoire the world leader in cocoa production is now causing desertification, drinking water shortages, a major loss of biodiversity, and many other problems. Some of the large agribusinesses recently consulted the Association Technique Internationale des Bois Tropicaux[28] for advice on how to avoid such problems; however, a 20 to 30 percent forest cover change still seems likely in the coming decades. Consumer pressure on agro-industry could be a valuable tool to reduce the deforestation.

Meanwhile, these are the country positions in Central Africa. The Central African Republic is in civil war, and in the Democratic Republic of the Congo, the government is still not in a position to control most of its national territory. These two countries will have difficulties adopting a consolidated position on forest conversion for agro-industrial plantations. Gabon[29] and the Republic of the Congo[30] do not seem to have a government policy on forest conversion. Only Cameroon has a slightly more advanced position.[31] Cameroon is directing agro-industrial plantations to the non-permanent forest domain that currently covers about seven million hectares. This domain is partly forested, so development within it will still result in some forest loss. According to certain reports, agro-industrial plantations are also being set up inside approved natural forest concessions, that is, inside the permanent forest domain.[32]

Thus, the problem of large-scale commercial agro-industrial activity is hitting the continent already. Large agro-plantations are being set up in Cameroon, the Republic of the Congo, the Democratic Republic of the Congo, and Gabon, but clear government policies are lacking and statistical information is hard to come by. The international donor community unfortunately remains somewhat indifferent and is not yet ready to invest in elaboration of policies and norms to help regulate forest conversion.

A new strategy to come up with solutions for competing land uses in remote rural areas is "landscape management." This means integrating competing land uses into a functional large area management plan for sustainable economic development. For now, nobody has a clear plan for this, but the Center for International Forestry Research is investigating such approaches[33] and the Association Technique Internationale des Bois Tropicaux is exploring solutions in the north of the

Republic of the Congo with a project that combines Forest Stewardship Council–certified timber concessions with protected areas supported by the Wildlife Conservation Society and local communities. But the Central African forests need far more attention in the years to come if the mistakes made in other regions are not to be repeated to the detriment of the forest, its people, and its biodiversity.

RALPH RIDDER *is the director general of the Association Technique Internationale des Bois Tropicaux (*ATIBT*) and the former director of the* FLEGT *and* REDD *Facilities of the European Union.*

Conclusions from regional deforestation trends

A CLOSER LOOK at the recent deforestation history in the three main tropical rainforest regions confirms that the dynamics of forest destruction are complex. While some large areas are cleared outright for commercial agriculture, many other deforestation hot spots simply disintegrate into a puzzle of smaller and smaller pieces of degraded, fragmented forests. In South and Central America over the past forty years, regional deforestation trends were largely overshadowed by the massive forest land conversion to commercial agriculture and cattle ranching in the largest tropical rainforest country, Brazil. Likewise, regional deforestation in the largest Southeast Asian country, Indonesia, dominates the regional statistics. A combination of commercial plantations (mainly oil palm) and small-scale subsistence agriculture have led to some of the highest deforestation rates ever recorded. The West and Central African deforestation, at least since 2000, hinges on very different circumstances. While West African countries had already lost a majority of their rainforests in previous decades, the largest forest country in the region, the Democratic Republic of the Congo, unexpectedly still has a moderating effect on the regional average. This effect can be credited to the remoteness of the areas within the Congo Basin and the fact that the commercial pressure for forest conversion has not reached it—yet. Small-scale subsistence agriculture, albeit largely on the heels of commercial logging, remained the main cause of deforestation, at least until 2010.

Tropical deforestation is not like the slow shrinking of a piece of ice in the sun; it is more like a crumbling of intact forest areas into variously degraded fragments of forest interspersed with agricultural or other nonforest land. The dynamic becomes even more complex if the inverse processes of natural regrowth and afforestation are happening at the same time. Satellite imagery of tropical rainforests shows these processes at work in all but the most remote areas. From space, these areas of deforestation outside the remote parts of rainforest (the deforestation hot spots) appear as a fine-grained patchwork that falls into various categories from closed forest to nonforest.

The complex dynamics of deforestation as forest areas move through stages of regrowth and reforestation are why we need standardized forest classification systems and satellite imagery interpretation. The lack of distinction between gross forest cover loss (the loss of the original vegetation cover, whether primary or secondary) and net forest cover loss (which includes reforestation, whether natural or planted) in the FRA 2010 statistics prevents us from truly understanding the dynamics at work.

A recent study scrutinized the current trend of governments, corporations, and nongovernmental organizations to set "zero deforestation" targets. It showed that unless separate targets were specified for reductions in gross deforestation on the one hand and reforestation on the other, such targets could be questionable, particularly at the global level.[34] The Brazilian National Institute for Space Research is to be commended for publishing the annual gross deforestation rates for the Amazônia Legal region as an exception to the general habit of reporting ambiguous net deforestation. The remote sensing specialists of the South Dakota State University and the University of Maryland have done an excellent job of demonstrating that such distinctions, including the separation between primary and secondary forests, are possible with modern remote sensing technology and interpretation. The high-resolution global forest cover change map with its multi-temporal pixel-by-pixel analysis of forest cover change has set a new standard that will have to become the norm, not least in Southeast Asia.

Finally, the finding that the majority of humid tropical deforestation occurs in specific deforestation hot spots, most clearly in some

Indonesian provinces as well as in the Amazonian arc of deforestation, shows that policies and incentive structures play a role in
deforestation patterns. The difference in the deforestation trends
that we see in Brazil and Indonesia (falling in Brazil and rising in
Indonesia) shown in figure 3.2 demonstrate that national or state
policies and incentive structures regarding things like land use planning, road building, prevention of illegal logging, and enforcement,
in combination with up-to-date satellite monitoring technology, can
bring solutions to the curse of seemingly uncontrollable tropical
rainforest loss.

JEFFREY SAYER

The Oil Palm: Threat to Conservation, Opportunity for Poor People

Rapidly expanding oil palm estates have destroyed vast areas of tropical rainforests in Asia, Africa, and South America in recent decades. Conservation activists who had previously seen unsustainable logging as the biggest threat to forests and their biodiversity have watched appalled as the relatively modest impacts of logging were totally overshadowed by whole forests being converted to monocultures of oil palm. Opposition to oil palm expansion has become the rallying ground for the conservation movement, with calls for moratoriums on new plantations and boycotts of oil palm products. Conversion of forests to industrial tree crops, of which the oil palm is by far the most important, has been a major cause of deforestation in recent years. But what has happened up until now may be eclipsed by some of the oil palm estates planned for the future.

The impacts of oil palm expansion merit examination from other perspectives. Many of the new estates are in countries where the people living in remote forest areas are among the poorest on the planet. Rural people in the rainforest countries of Africa mostly struggle to survive on unproductive subsistence agriculture. Even in Indonesia many people who live in the areas affected by oil palm plantations live below the poverty line and have few economic options for improving their lives. And in South America the oil palm is expanding in areas where the least prosperous sectors of the population live. We have to examine the effects of palm oil production from the perspectives of these people—after all, the oil palm is being planted on their land.

So, is the oil palm an inherently bad crop? Absolutely not. The oil palm is one of the most profitable crops that can be grown on the poor soils of the humid tropics. It produces more value per hectare than any other industrial crop. It also has a better carbon balance than other crops that might be grown in the same areas, and it survives with less pesticide and fertilizer use than other crops. The oil palm also yields fruit throughout the year and requires a lot of unskilled labor. So if rainforests have to be replaced by anything, then the oil palm may be the least bad option. The oil palm has the added advantage that it is very well adapted to smallholder farmers. A family farm of two to ten hectares can be highly profitable, and smallholders often combine oil palms with other tree crops and basics such as rice or maize and vegetable crops. Landscapes covered in these small farms or gardens are diverse, and if small pockets of natural forests survive along streams or on hillsides, they support quite a lot of biodiversity. In parts of Nigeria, Indonesia, and Malaysia, smallholder oil palm landscapes meet many of the criteria for sustainable development in the humid tropics.

The problem then is not that the oil palm is a bad crop—it is the way that the recent expansion of oil palm plantations has occurred that has provoked the wrath of environmentalists.[1] The oil palm is a very profitable crop, but only under certain conditions. One of them is that a processing plant needs to be accessible near the location where the palms are grown. Palm oil mills require major investments that are way beyond the means of the rural poor. Technological advances mean that palm oil now has a multitude of uses including cosmetics, soaps, edible oils, and biofuels. Palm oil mills have become attractive targets for major corporate investors, and these investors want to be sure of their sources of raw materials.

This combination of factors has led to a proliferation of large-scale oil palm initiatives throughout the humid tropics. Much of this expansion has occurred in processes where government oversight is weak and local land rights are not easily defended. Poor rural people have found their land seized by corporations and converted to oil palm estates. To make matters worse, the corporations often import their own labor from other areas, and local people get few benefits. An added issue has been that the oil palm companies recover the timber from the land that they clear, and they get access to this timber without having to respect the rules of sustainable forestry that would apply in areas allocated for forestry. The timber is a waste product—but a valuable one.

Oil palm expansion has been the main driver of deforestation in some of the world's most precious forest landscapes. Corporate investors have become rich, and local people have lost their land. To confront these complex and difficult challenges, a Roundtable on Sustainable Palm Oil (RSPO) was established by conservation organizations and some of the more farsighted industrial investors (see box 4.3). In some countries the RSPO standards are now backed up at least to a limited extent by government planning agencies. The RSPO, however, has no authority and cannot take action against companies that fail to respect its standards. Some environmentalists complain that the RSPO has no teeth, whereas industry contends that environmentalists fail to recognize the economic benefits that flow from the industry.

Oil palm expansion continues to present great challenges. The vast development schemes for eastern Indonesia that were recently announced appear to have bypassed national regulatory measures and ignore the RSPO. But progress is being made with programs that combine industrial estates with smallholder producers. They are increasingly favored in national policies—although not everywhere— and many larger companies are taking their environmental and social responsibilities more seriously. The RSPO has provided a forum where civil society has been able to confront the industry and where issues have been critically reviewed. There is still much to be done, but we at least know what has to be done if oil palm growing is to play its role in driving economic growth, alleviating rural poverty, and contributing to sustainable development in poor tropical countries.

JEFFREY SAYER *is a professor at the School of Earth and Environmental Sciences, James Cook University, in Cairns, Australia. He is a former director general of the Center for International Forestry Research* (CIFOR).

Wood engraving from the middle of the nineteenth century depicting indigenous people in the rainforests of the Nicobar Islands.

4

THE ROOT CAUSES OF DEFORESTATION AND FOREST DEGRADATION

F TROPICAL DEFORESTATION is to be stopped or at least reduced, it is essential to understand what drives deforestation. New strategies such as REDD+, which will offer economic incentives for avoided deforestation, will depend on understanding the root causes of forest destruction. Although poverty and commercial interests are most often seen as driving forest loss, not all land use decisions that lead to deforestation are economic. Sociocultural causes can also contribute to deforestation through a combination of economic, political, and institutional factors aimed at gaining additional agricultural area. In Brazil, for example, the Trans-Amazonian road construction scheme of the 1970s was accompanied by colonization programs, fiscal incentives, and expansion of cattle ranches which later became the main drivers of deforestation.[1]

Direct drivers and their underlying causes

THE FACTORS THAT drive deforestation vary greatly over different places and times. Since forest areas may be cleared or degraded for one use in the first place and later converted to another use, it is not always possible to relate a specific deforestation driver to a corresponding land use type. For example, often road construction that goes with infrastructure development or timber extraction later leads to agricultural expansion. Until the early 1990s the two most often mentioned direct causes of tropical deforestation were shifting cultivation in combination with population growth, and timber exploitation—logging.

Two researchers at the University of Leuven in Belgium analyzed 152 case studies of tropical deforestation, considering direct causes on the one hand and the underlying driving forces on the other.[2] Not surprisingly they found that the picture is diverse, with a number of factors driving several direct causes of deforestation. However, now and in the past, agricultural expansion is by far the most important land use change associated with deforestation. Agricultural expansion in tropical rainforest areas falls into two main categories: small-scale shifting cultivation and subsistence farming on the one hand, and large-scale commercial plantations and pastures on the other. Of the more than 100 million hectares (net increase) of agricultural land in the tropics from 1980 to 2000, more than 55 percent came from intact rainforests and about 28 percent from degraded rainforests.[3]

In recent years, concern about carbon emissions from tropical deforestation, as well as the development of REDD+, has spurred investigation of what drives deforestation.[4] These studies confirm that commercial agriculture, including cattle ranching, is now the most important proximate driver of deforestation and has overtaken subsistence agriculture everywhere except in Africa.

FIG. 4.1. Forest area changes

Proportion of forest area changed (A) and forest area degraded (B) in 2000–2010 by proximate (direct) drivers. Based on data from forty-six countries, which together cover 81 percent of the forest loss during this period in tropical and subtropical countries. The data include some areas outside the tropical rainforest biome, particularly for Africa. Adapted from Hosonuma et al. (2012).[5]

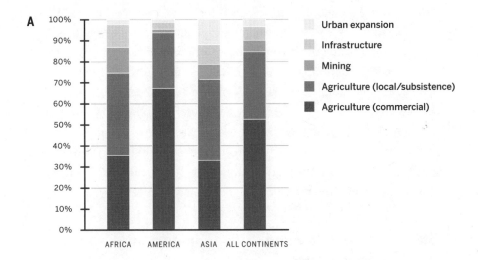

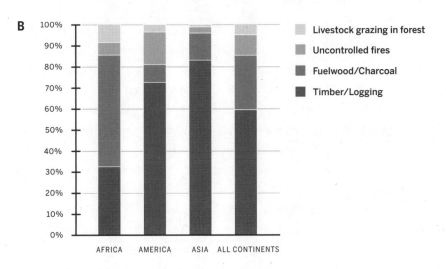

Rainforest areas that are converted to agricultural land or pastures often do not remain under permanent land use. Because of the low fertility of rainforest soils, cleared rainforest areas are commonly abandoned after a few years. Thus, clearing forest land for agriculture does not produce an equivalent area of agricultural land: it produces agricultural land plus large areas of unproductive fallow, farm bush, and degraded pasture, as can be seen in many tropical countries. Forest conversion for agriculture is an erosive, land-consuming process, and it comes mostly at the cost of intact forests.

Forest degradation, on the other hand, is still primarily driven by timber exploitation, particularly illegal exploitation (see figure 4.1). Before analyzing the "modern" trends of deforestation, driven by commercial agriculture, we will first look at some of the more traditional causes of tropical deforestation—subsistence farming practiced by shifting cultivation on the one hand and selective logging as the main driver of forest degradation on the other.

Slash-and-burn farming: Still an important deforestation driver

TRADITIONAL SHIFTING CULTIVATION, also called "swidden agriculture," is an old form of subsistence farming where periods of cultivation are followed by the land lying fallow and ultimately developing into closed forest again. It is still practiced by traditional forest communities and indigenous peoples. Today, slash-and-burn farming, where larger areas are cleared, is more common. Slash-and-burn farming, often used to produce a mixture of food crops and cash crops, is mostly practiced by immigrant settlers and leads to permanent land conversion.

Central Africa differs from other regions in that although deforestation is common, it remains primarily driven by small-scale farmers and less by large-scale commercial agriculture. The deforestation dynamics in Africa, however, are not well understood, and Central Africa could simply be a latecomer to the large-scale forest cover changes we see in Southeast Asia and South America.

Since the 1990s, remote sensing of Central African forests documented not only lower deforestation rates than in other continents, but also a decline in deforestation rates in the twenty-first century (see figure 3.2). These trends have been attributed to the virtual absence of forest conversion for large-scale commercial agriculture. Another theory maintained that because the Congo Basin countries are rich in oil and mineral reserves, income from extractive industries promoted urbanization and made small-scale agriculture in remote areas unattractive.[6] The effect of revenue from the petroleum industry undoubtedly applies to Gabon with its low population and relative prosperity.

But the situation is very different in other countries, notably the Democratic Republic of the Congo. In fact the Democratic Republic of the Congo, but also the neighboring Republic of the Congo, showed signs of an inverse trend—an acceleration of deforestation in the 2005–2010 period. This increase was mainly due to subsistence agriculture in the more densely populated areas, particularly in the eastern provinces of the Democratic Republic of the Congo (see plate 6). The influx of immigrant farmers and people displaced by war is taking a heavy toll on the forest in this part of the Congo Basin. Local charcoal production and mining also contribute to the increase in deforestation.[7] A particularly disturbing increase in deforestation in the Democratic Republic of the Congo is linked to the consumption of wood fuels, mainly charcoal.

SHIFTING CULTIVATION AND THE NEED FOR WOOD FUELS DRIVES DEFORESTATION NEAR URBAN AREAS

HOUSEHOLDS IN KINSHASA and Kisangani, Democratic Republic of the Congo, depend on wood fuels for more than 85 percent of their energy needs, but breweries, brickmakers, and aluminum smelters also use large quantities. The volume of wood fuels consumed by Kinshasa and Kisangani alone is more than twelve times the national timber production.[8] The pattern of deforestation hot spots (see plate 6) reveals that deforestation is largely determined by proximity to urban settlements and roads. The opening up of remote areas by timber concessions that can only be accessed on long-distance timber roads, on the other hand, does not lead to higher deforestation. Beyond a distance of nine to twelve hours of travel time from a city (not very far with the means of transport available), deforestation in the Congo Basin drops to almost zero,[9] showing that the theory that timber extraction always leads to deforestation by slash-and-burn farming does not hold true for remote forest areas in the Congo Basin.

The reason why slash-and-burn farming is less attractive in remote areas is simple. Small-scale farmers do not live in a cash-free economy, so they rely on a mixture of food crops and cash crops. Slash-and-burn farming can only survive where farmers have access to local markets to buy their own supplies of sugar, salt, soap, cooking oil, and other commodities and to sell their cash crops. Small-scale farmers tend to invade forest areas along main road corridors and along the Congo River and its navigable tributaries, while staying close enough to towns where their produce can be traded. This is why the interiors of the Congo Basin have thus far remained relatively safe from deforestation by shifting cultivation, despite extensive timber concessions. But the writing is on the wall that this may change in the not-too-distant future.

Selective logging: The good, the bad, and the ugly

MOST PEOPLE EQUATE the destruction of tropical rainforests with logging—after all, the terms for deforestation in German and French, "Abholzung" and "déboisement," both refer to the removal of wood. These terms imply that logging is the prime driver of deforestation.* Until a few decades ago, tropical hardwood extraction did indeed dominate the commercial interest in tropical rainforests, but it focused on very few of the hundreds of tree species, and within these commercially valuable species, only trees of the right quality and diameter would be selected for felling.

In most rainforest areas the large majority of tree species have no commercial value as timber, either because they are unsuitable as sawn wood or plywood, or because the species is unknown or untested. Buyers of tropical timber are notoriously unwilling to experiment with lesser-known species for fear of costly surprises. On the wood market these species find hardly any customers, and as a result, tropical forestry remains highly selective. For this reason as well as legal reasons, tropical forestry rarely involves clear-cutting—fortunately.

WHY SOUTHEAST ASIAN LOGGING IS OFTEN FATAL TO THE FOREST

THE FORESTS OF Southeast Asia, however, are an exception. Commercially valuable timber species are more abundant, and their exploitation often leads to complete deforestation. In Southeast Asian lowland rainforests, the Dipterocarpaceae are particularly well represented, with many species of commercially valuable trees, including the group of meranti timber species. Because these forests contain a high density of valuable dipterocarps as well as the ramin timber species (*Gonystylus sp.*), they are much more intensively logged. With conventional logging practices, often more than 50 percent of the stand is extracted or damaged, which makes it unlikely that forests will recover enough to produce a reasonable yield even after a full felling cycle of thirty-five years. Their low

---

* Foresters and timber companies usually prefer the term "timber harvesting," but the vernacular term "logging" is now also commonly used for the whole process of planning, felling, extraction, and processing timber logs. "Timber extraction" on the other hand has a narrower meaning of transporting of logs from the felling site.

economic value after the first cut also makes over-logged forests more likely to be clear-cut and converted to agricultural land.[10]

Revenue from timber exploitation in Southeast Asia is often used to finance oil palm farming, which has been the principal deforestation driver for the last few decades. Commercial wood extraction was linked to more than three-quarters of cases of deforestation in this region.[11] Because unsustainable logging has depleted timber volumes or caused secondary forests to be converted to oil palm plantations, Southeast Asian companies are looking to Central Africa and South American countries (Brazil, Guyana, and Suriname), as well as to South Pacific Islands (Solomon Islands and Papua New Guinea) for new timber sources.[12]

SELECTIVE LOGGING IN AMAZONIA AND CENTRAL AFRICA

IN AMAZONIAN AND African tropical rainforests timber extraction is far more selective than in Southeast Asia because of the lower density of commercial timber species. In Amazonia, only one to three trees per hectare are removed in newly logged areas, although more intensive harvesting may take place in previously logged areas. Most of the impacts of logging are caused by the forest roads, skidder tracks, and loading areas that are necessary for extracting trees. The new roads also allow migrant settlers who live from shifting agriculture to move into the area.[13]

Although the diversity of tree species per hectare in African rainforests is comparable to that of Southeast Asian and South American rainforests, African rainforests harbor an even lower density of commercial species. Therefore timber exploitation in African rainforests is particularly selective. Often not more than one tree per hectare is harvested in a felling cycle of twenty-five or thirty years. Not surprisingly, the Central African timber yield makes up only 0.4 percent of the world's roundwood production. But highly selective exploitation also means that larger areas of forest are affected (see plate 3). Timber concessions in Central Africa cover about forty-four million hectares—about a quarter of the lowland rain-forest area, larger than the state of California. The size of this area has hardly changed over the past years, except in the Democratic Republic of the Congo, where it has actually shrunk since 2002.[14]

A recent study of the drivers of deforestation in African rainfor-ests confirmed that small-scale farming and increasing demand for fuelwood are key drivers, whereas timber exploitation, somewhat surprisingly, was found to have little or no effect.[15] Part of the reason logging concessions do not contribute more to deforestation may be due to efforts like those of the Central African Forest Commission, COMIFAC,[16] which requires Central African countries to introduce management plans for timber concessions.[17] Starting with Cameroon, sustainable forest management has increasingly replaced logging without management plans since the start of the new millennium and is now the prevailing trend in Central Africa. About fourteen mil-lion hectares (about one-third of the timber concession area) were under formal management in 2010, and FSC certification reached 4.6 million hectares in the Congo Basin in early 2014 (see box 4.1).

Sustainable forest management, particularly if FSC certified, and low-impact logging practices provide a good balance of a reasonable revenue from forestry operations with employment in rural areas. Sustainable forest management is (slowly) becoming more wide-spread, but the forestry sector has been facing harsh times since the financial crisis in 2008–2009 when the world tropical timber trade shrank by one-third and prices suddenly fell by 15–30 percent. In Central Africa alone, 25,000–30,000 jobs were suspended and some companies shut down.[18] Although some of the radical non-governmental organizations celebrated these developments, they are actually bad news for the Congo Basin forests. This vacuum may be rapidly filled by small or illegal operators that fall short of any sustainability standards. Worse, the incentives for more profit-able agribusiness on converted forest land are increasing in Central Africa (see box 3.1).

BOX 4.1   **The FSC: Losing ground in the tropics**

The FSC (Forest Stewardship Council) was founded in 1993 to ensure environmental, social, and economic standards for sustainable forest management based on agreed principles and a product labeling system that guarantees product recognition on the market. The FSC certification system monitors compliance against the standard through accredited certifiers. Thus, the FSC is a performance standard, which distinguishes it from other forestry standards without control systems. The FSC has grown enormously, particularly since 2000. In early 2014 it reached 182 million hectares of certified forest area worldwide, equivalent to the combined area of Germany, France, Spain, and Italy!

When the FSC was established, soon after the Earth Summit in Rio, it focused on sustainable forest management in the tropics. It is disappointing that today only 11 percent of the globally certified forest areas are in the tropics and subtropics. Tropical forest companies that produce FSC certified timber face great difficulties: the cost of certification, competition from illegal logging, and even attacks from radical groups. A number of them have recently sold their operations, and FSC certification has therefore lost ground to weaker certification schemes and unsustainable and illegal logging. The FSC would do well to put far more effort into promoting FSC certification in the tropics and promoting their products in consumer markets.

Nongovernmental organizations are crucial for promoting FSC certification in the tropics, and they should see certification as a viable method to prevent or decrease forest conversion to other land use. The Global Forest and Trade Network run by the WWF is a network of companies involved in harvesting, processing, and trading timber. The WWF encourages member companies to increase the proportion of their timber that comes from verified sources. The trade network helps its member companies move from using unknown sources through a known licensed source (legal origin) to a fully certified source by linking producer members with consumers.

THE PROSPECTS FOR SUSTAINABLE LOGGING

THE PREVALENCE OF selective logging in tropical rainforests has been used as a justification by the timber industry, which commonly claims that "selective" also means "sustainable." Although that is a blatant oversimplification, it is true that selective timber extraction is rarely a direct driver of deforestation, at least not in Africa and Latin America.[19] However, it causes varying degrees of degradation in a very large part of the world's rainforests. Between 2000 and 2005, about 28 percent of the tropical rainforest in Asia and Oceania was under concessions for selective logging, 20 percent in Africa, and 18 percent in South America.[20] This makes up an estimated area of 398 million hectares, an area considerably larger than India.

Because such a vast forest area is affected by forestry operations, sustainable forest management is of the utmost importance. In theory tropical timber harvesting can be a tool to give forests sustainable value, but selective harvesting is not a sufficient criterion—far from it. Selective logging can take many forms, from carefully planned low-impact logging that leaves hardly any trace after a felling cycle of thirty years to the sadly devastated rainforest areas seen in the Malaysian state of Sarawak and other Southeast Asian forest areas. Extracting even a single tree per hectare can create a lot of "collateral" damage from other trees being damaged or pulled down by the falling timber tree, as well as from skidder tracks and loading areas.

Low-impact or reduced-impact logging requires the operator to carefully select the tree to be felled, cut the vines and lianas, plan the skid tracks so they do the least damage to the forest soil and other trees, and keep loading areas as small as possible (see figure 4.2). Under optimal circumstances such timber exploitation does not leave larger gaps in the canopy than the natural collapse of an emergent tree would cause. Depending on the circumstances, a low-impact extraction may leave almost no trace after ten or twenty years. In fact such practices were already being applied in

West Africa fifty years ago.[21] I have visited such forest concessions in Africa twenty or thirty years after the first cut and found that some of them could not be distinguished from a primary forest. The careful removal of one or two trees per hectare after a few decades was not important compared to the other consequences of timber extraction (see plate 4).

FIG. 4.2. Old logging road in Ghana

In African rainforests often no more than one to two trees per hectare are cut in a felling cycle every twenty-five to thirty years. The African umbrella tree (*Musanga cecropioides*), a fast-growing secondary forest species, overgrows an old forest road in western Ghana. (Photo by C. Martin.)

Logging has a reputation for preceding slash-and-burn farming. Selectively logged areas are more likely to be affected by shifting cultivation because timber roads provide easy access to forest areas, particularly near towns and cities, as the case of logging in the Congo Basin demonstrates. Easy access to the forest leads to increased hunting pressure and forest fragmentation through slash-and-burn farming and edge effects. These consequences are much more serious than the removal of a few trees.

Therefore, to protect biodiversity we should pay far more attention to the impact of forest fragmentation, hunting, and the bushmeat trade than to whether cutting a given tree could threaten the timber species in question. Certain wildlife species, such as forest elephants, gorillas, and duikers, have been found to prefer logged-over forest because more gap vegetation and vines are available. Secondary forests can still be valuable for biodiversity conservation, and they can enlarge the conservation estate of protected areas, as research on logging concessions adjacent to the Nouabalé-Ndoki National Park in the northern part of the Republic of the Congo has shown.[22] I am by no means suggesting that more primary forests be opened for logging, but I am saying we should recognize the biodiversity value of secondary forests and mitigate the devastating effects of overhunting in such areas.

The close relationship between illegal logging and corruption

WHEN IT COMPLIES with forestry laws, timber harvesting has not been found to contribute significantly to deforestation in most tropical rainforest countries. It does, however, cause forest degradation to various degrees, depending on whether or not forestry operates sustainably. In 2014 only about one-third of the area recognized as production forest by the International Tropical Timber Organization came under approved management plans, and only 6.5 percent was certified. Although an increasing number of countries require approved management plans for timber operators, this leaves a lot of room for unregulated and uncontrolled logging activities.

Not surprisingly illegal logging[23] is a very serious problem that contributes in a major way to deforestation. Since the first decade of the new millennium it has triggered a number of mitigation campaigns. A detailed study of illegal logging in five tropical countries (Indonesia, Malaysia, Ghana, Cameroon, and Brazil) published by Chatham House in 2010 found that illegal logging has been declining since about 2004. But at the time this study was published, illegal harvesting still represented 35–72 percent of all logging in the Brazilian Amazon, 22–35 percent in Cameroon, 59–65 percent in Ghana, 40–61 percent in Indonesia, and 14–25 percent in Malaysia.

Illegal logging affects not only tropical countries, but industrialized countries as well. Worldwide, more than one hundred million cubic meters of timber are still being cut illegally each year, leading to the degradation or destruction of five million hectares of forest.[24] The economic damage of illegal logging is equally important; it is estimated at USD 10–15 billion worth of stolen timber per year, plus an additional USD 5 billion in lost royalties and taxes.[25]

THE IMPORTANCE OF CRIMINAL JUSTICE SYSTEMS

TWO IMPORTANT AND often interlinked factors have been identified as underlying causes of illegal logging: corruption, especially at higher levels of government, and failing criminal justice systems.[26] In Indonesia an initiative to combat illegal logging in the province of Papua in 2005 found 186 suspects and confiscated 400,000 cubic meters of timber. But ultimately only thirteen offenders were convicted. When the Indonesian Anti-Mafia Law Task Force revised illegal logging court cases in 2010, it found that of ninety-two accused offenders, forty-nine were acquitted and the rest received short jail terms.

These too-lenient sentences, often by corrupt judges, are not enough to deter profitable criminal logging activities. A case of illegal logging in the Brazilian state of Mato Grosso revealed the link between corrupt government officials and illegal logging. In July 2010 the Brazilian federal police under Operation Jurupari uncovered a criminal scheme involving top officials in the administration of the governor of Mato Grosso as well as other officials in Brazil's environment agency, which oversees the logging industry. The illegal logging scheme involved an estimated 1.5 million cubic meters of illegally harvested timber and an estimated USD 500 million in damage.[27]

FIG. 4.3. Relationship between illegal logging and corruption in Indonesia

Between 2000 and 2006 the percentage of illegal log supply decreased with lower levels of corruption. The level of corruption is based on the Transparency International Corruption Perceptions Index. Adapted from a study by Chatham House.[28]

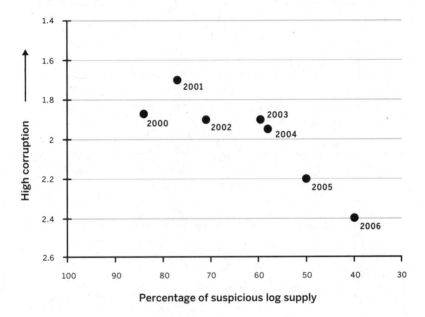

Percentage of suspicious log supply

However, both Indonesia and Brazil have demonstrated that political will and resolute government action against corruption can make a difference. The decline of illegal logging in Indonesia was a consequence of the "Presidential Instruction on Illegal Logging" in 2005 and extensive enforcement measures (see figure 4.3). International publicity, nongovernmental organization investigations, and intergovernmental schemes against illegal logging are also effective. Since the mid-2000s considerable reductions have been achieved in a number of tropical countries. The target countries of the Chatham House study reduced illegal logging by more than 50 percent, although much remains to be done (see figure 4.4). China, for example, compensated for decreased illegal wood imports from Indonesia by importing more illegal wood from Papua New Guinea (see box 4.2).

FIG. 4.4. Imports of illegally sourced wood and wood products in the largest importer countries

After 2004 the imports of illegally sourced wood by China started to decrease because of the decline in illegal logging in Indonesia. These reductions have been partly offset by increased imports of illegal timber from Russia and Papua New Guinea. Imports of illegally sourced timber from Indonesia into Japan and the United States declined as well. In the United States the decline of illegal timber importation started after some delay in 2007. This was due to the continued increase of imports of illegally sourced timber in cheap wood products (furniture) from the main processing country, China. The fall of imports of illegally sourced timber by the United States in 2008 must be attributed to the economic slowdown. RWE means roundwood equivalent (see Glossary). Adapted from Lawson and Macfaul.[29]

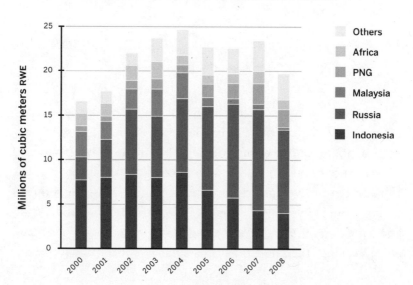

ESTIMATED ILLEGAL WOOD IMPORTS BY CHINA

ESTIMATED ILLEGAL WOOD IMPORTS BY JAPAN

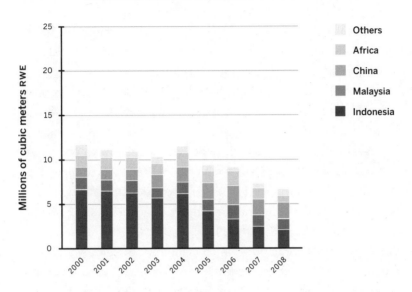

ESTIMATED ILLEGAL WOOD IMPORTS BY THE UNITED STATES

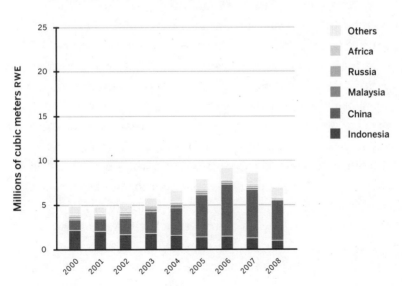

INTERNATIONAL COOPERATION AGAINST ILLEGAL LOGGING

THE INTERNATIONAL COMMUNITY has responded to illegal logging by introducing regulations and control mechanisms. In the European Union the Forest Law Enforcement, Governance and Trade (FLEGT) action plan consists of two key elements:

1) The FLEGT Regulation, adopted in 2005, provides a framework for countries to enter into bilateral voluntary partnership agreements (VPAS) in which both parties agree to verify that exported timber is legal. As of early 2014, VPAS have been concluded with Ghana, are under ratification with Cameroon and the Republic of the Congo, and are in negotiation with Liberia, Gabon, the Democratic Republic of the Congo, the Central African Republic, Malaysia, Indonesia, and Vietnam.

2) The EU Timber Regulation, which came into force in March 2013, prohibits the "placing on the [EU] market of illegally harvested timber or timber products derived from such timber." One year into the introduction of the timber regulation, its track record is mixed. Many governments have not committed to effective enforcement, and the "competent authorities" they were obliged to put in place are often less than competent or lack the necessary resources.

In 2008 US Congress followed suit and passed an amendment to the Lacey Act of 1900, which prohibits trafficking in illegally harvested plants and wildlife. The act now includes illegally harvested wood and wood products and is being enforced by the US Fish and Wildlife Service. The Lacey Act has already proven its effectiveness with a widely reported case in 2012 involving the Gibson Guitar Company. Gibson was charged with importing illegally logged timber from Madagascar's rainforests in 2008 and 2009. The company was fined USD 300,000 for violating the Lacey Act and had to pay USD 50,000 to the National Fish and Wildlife Foundation to promote the conservation of protected tree species used in musical instruments.[30]

In 2012 the Australian government also passed new legislation— the Illegal Logging Prohibition Act. Legal frameworks to contain illegal timber imports also need to be introduced in the large Asian consumer countries, Japan and China. As a major timber importer for its own consumption as well as timber processing, China is currently piloting a timber legality scheme in cooperation with Australia, Japan, and Indonesia.

WHY ANTI-CORRUPTION MEASURES ARE ESSENTIAL TO FUTURE PROGRESS

THE MEASURES UNDERTAKEN by national governments, international organizations, nongovernmental organizations, and Chatham House[31] since the beginning of this millennium are encouraging, but the problem of illegal logging will require further effort, and illegal logging cannot be solved without addressing corruption.

Since illegal logging is a major driver of deforestation, instruments such as REDD+, which will provide compensation for avoided deforestation, cannot be effective unless the problem of corruption is resolved. Those involved in illegal logging will not be the beneficiaries, and since the gains from illegal logging activities are likely to exceed the REDD+ incentives, REDD+ readiness must mean addressing corruption first. The REDD+ program certainly discusses corruption, but mostly in relation to distributing incentives. REDD+ is unlikely to succeed in countries where corruption still provides a fertile ground for illegal logging activities.

Another problem is that an increasing proportion of illegal timber is not exported but is consumed within the countries themselves. The Chatham House study found that the large majority of illegal timber comes from small-scale artisanal logging, where corruption is particularly difficult to control.[32]

The footprint of rising consumption

SHIFTING CULTIVATION BY small farmers remained by far the most important cause of tropical deforestation until after World War II. Because European markets were relatively close to West Africa, commercial use of rainforest areas had started already in the nineteenth century. Europe's trade relations with the West African coast in fact date back to the fifteenth century, when the trade of gold, ivory, and kola nuts did not require clearing the forest. In the latter part of the nineteenth century, gum copal collected from *Daniella* trees for making varnish and a local rubber tapped from *Funtumia* trees were exported in increasing quantities, and by 1884 the first successful oil palm plantation in southern Ghana had exported up to 30,000 metric tons of palm oil. Ghana and Nigeria converted increasing stretches of their coastal rainforest areas to cocoa plantations in the last quarter of the nineteenth century.[33]

However, it was not until the 1950s that large plantations of rubber (*Hevea brasiliensis*) and oil palm plantations were established and operated at a profit in West Africa. Ghana became the largest cocoa producer in the world until it was overtaken by Côte d'Ivoire in the 1970s. West Africa thus ironically became the main beneficiary of a South American fruit tree, the cocoa tree (*Theobroma cacao*), while it was outcompeted more than fifteenfold by Southeast Asia in the production of palm oil from its native oil palm (*Elaeis guineensis*).

The conversion of tropical rainforests for larger-scale plantations is a relatively recent phenomenon. Technical obstacles to clear-cutting, such as lack of roads and heavy machinery, as well as the limited market for such agricultural produce as cocoa and coffee prevented large-scale deforestation for commercial crops until the 1960s and 70s in practically all rainforest areas of the world. Until that time, shifting cultivation worldwide remained the most important cause of deforestation.

PLANTED FORESTS INCREASINGLY REPLACING NATURAL FOREST

UNLIKE THE CONVERSION of natural forests for agricultural use, be it cocoa, coffee, oil palms, or soybeans, the FAO classifies conversion to tree plantations, for example teak, eucalyptus, rubber trees (Hevea), or pulp and paper plantations not as a loss of forest area, but as a change to another forest land use. On a global scale the conversion of natural tropical forests to planted forests, although increasing in many countries, is a less important driver by far for natural forest loss than the conversion to agricultural land.

According to FRA 2010 statistics, planted forests made up the highest percentage of the total forest area in Southeast Asia compared to other regions: 21 percent in Thailand, 9 percent in Malaysia, and 4 percent in Indonesia. Some West African countries, Côte d'Ivoire, Ghana, and Nigeria, follow with 3–5 percent. A large part of these plantations is made up of rubber trees (Hevea brasiliensis). Some Central American and Caribbean countries also have a high percentage of planted forests: 9 percent in Costa Rica, 5 percent in Guatemala, and 17 percent in Cuba. The other rainforest countries, including all of the large Central African and South American ones, have less than 2 percent of their forest area under planted forests.

RAINFOREST BIODIVERSITY TURNED INTO PULP AND PAPER

THE PRODUCTION OF tropical hardwoods has been decreasing since 2000 in a number of formerly important producer countries, notably Indonesia and Malaysia, and some Central African countries, not least the Democratic Republic of the Congo. At the same time the demand for pulp and paper is rising very rapidly. Pulp and paper production is now one of the largest industrial sectors in the world, with a rapid increase in consumption in the emerging Asian economies. The United States, Canada, and China account for more than 50 percent of world production, but the industry is moving south to where production costs are lower.[34]

Among the rainforest countries Brazil and Indonesia stand out with about ten million metric tons of annual pulp wood production each in 2011. Because of the huge forest area of Brazil, this production affects only about 1.4 percent of its forest area; it also has not increased since 2000. The largest rainforest conversion for pulp and paper production occurred in Indonesia, particularly in Sumatra, where it has contributed in a major way to deforestation.

In 2011 a coalition of nongovernmental organizations including WWF Indonesia published the report "The Truth Behind APP's Greenwash," which claimed that logging companies supplying Asia Pulp and Paper (APP) (which is owned by the Sinar Mas Group; see page 91) have destroyed two million hectares of Sumatran forest in the Riau province since 1984.[35] Six months later APP published a roadmap to 2020 that committed the company to sourcing its raw materials exclusively from tree plantations.[36] The second-largest pulp and paper producer, APRIL (Asia Pacific Resources International Limited), followed suit with a similar commitment in 2013.[37]

The massive Asian demand for paper and cardboard products, however, makes it extremely likely that the expansion of pulp and paper production in Indonesia will continue to encroach on natural forest areas (see plate 7) and that the companies involved will reach out to other countries, for example to Papua New Guinea (see box 4.2), or even such vulnerable continents as Africa. The jury is still out on the sustainability of pulp and paper production in Indonesia and elsewhere, and governments and nongovernmental organizations will have to remain extremely vigilant.

BOX 4.2   **Papua New Guinea: The new illegal logging frontier**

When Indonesia cracked down on illegal logging, some of the companies involved shifted their activities to other countries. After China
began to import less illegally sourced timber from Indonesia after 2005,
it partly made up the shortfall with illegally sourced timber from Papua
New Guinea. A recent study by the Oakland Institute[38] found a veritable
rush of foreign corporations on the forests of Papua New Guinea taking
place under the Special Agriculture and Business Leases scheme. An
area of 5.5 million hectares has been leased to foreign corporations in
recent years, in addition to 8.5 million hectares that had been allocated
earlier as logging concessions. Although this scheme was launched to
promote agricultural projects, these corporations seem to be primarily
interested in lucrative short-term gains from log exports. Illegal logging
that thrives on corruption, poor governance, and lack of enforcement
dominates the timber sector. The largest logging companies in Papua
New Guinea are of Malaysian origin, the Rimbunan Hijau group among
them, but a commission of inquiry found that companies from Australia, the United States, and some Asian countries were also involved in
irregular land deals.

Papua New Guinea has some of the richest biological and cultural diversity in the world. Linguists have recognized 850 indigenous
languages, which makes the country the most linguistically diverse
country in the world by far. In principle the country's constitution guarantees the local tribes' customary rights on their land, but today about
30 percent of the land area—equivalent to twice the size of Ireland—
has been leased to foreign corporations. By 2021 about 83 percent of
the accessible forest areas may be deforested or severely damaged,
according to the Papua New Guinea Forest Authority.

The deforestation and the biological as well as cultural degradation
in Papua New Guinea raises the legitimate question of why such an ill-
fated development should still be possible today, after the lessons we

have learned from rainforest destruction in Indonesia and Latin America in the past decades. The Oakland Institute's study concludes that: "The key problem is a development agenda based on unrestrained capitalism, foreign investment, and resource exploitation—operating within a context of widespread corruption and a dysfunctional administration."

The global importance of Papua New Guinea's forests and cultures requires urgent action by a responsible government and international cooperation of governmental and intergovernmental institutions, non-governmental organizations, and enforcement mechanisms such as Interpol. Only then can one of the worst environmental and cultural disasters of our time be prevented.

The oil palm: An African descendant wreaks havoc in Southeast Asia

WHOEVER SAW THE magnificent oil palm (*Elaeis guineensis*) in its natural habitat in the rainforests of western Africa a few decades ago would never have believed that this plant could ever become a threat to the biodiversity of faraway continents. Perhaps one may be forgiven for not having taken human ingenuity and greed into account. Some of the first oil palm plantations were established in Ghana around 1850. The edible oil, known for centuries to forest dwellers, was a blessing for the local economy. But even when the first oil palm plantations were established in Peninsular Malaysia almost a hundred years ago, this did not look like a threat to African producers. Malaysian production was less than satisfactory, as the palm trees had to be pollinated manually—a tedious and inefficient process.

The oil palm was believed at that time to be wind-pollinated, until R. A. Syed from the Commonwealth Institute of Biological Control in Rawalpindi, Pakistan, found an indigenous insect in the Pamol (Unilever) plantation in Cameroon that was an effective pollinator—a small weevil (*Elaeidobius kamerunicus*). When the weevil was introduced in Malaysian plantations in 1981, the yield increased by 40–60 percent and Malaysia and Indonesia started to dominate the world trade in palm oil.[39] Ironically, the Pamol plantation in Cameroon, where the "gold bug" was discovered, went bankrupt a few years later, in 1987.[40]

FIG. 4.5. Production and consumption of palm oil by country in 2011–12

Asian countries, including the two main producer countries, Malaysia and Indonesia, are the main consumers of palm oil. European countries have also become important consumers. Source: AOCS (American Oil Chemists' Society).[41]

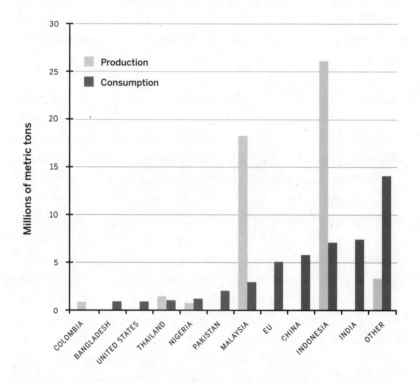

Palm oil is derived by pressing it from the fibrous pulp of the reddish fruit, which gives it its red color. The oil from the palm kernel can also be used for food and is traded as an industrial oil. Palm oil is widely used for cooking in developing countries, mainly in Asian countries, and the world's most important consumer country is India (see figure 4.5). Today, palm oil is everywhere; besides being used as cooking oil, it is used in thousands of products in the food industry and also in soap, detergent, lubricants, cosmetics, and as biofuel.[42]

It makes up the largest volume of vegetable oil produced, at 34 percent, and palm kernel oil adds another 4 percent to the global palm oil bounty.[43] It is the cheapest edible oil and by far the most productive: plantations yield three to eight times as much oil by area than any other oil plant.[44] Currently more than 70 percent of the world's palm oil is used for food. The main producers, Indonesia and Malaysia, in addition to their exports, meet a large part of their domestic demand for vegetable oils with palm oil, but there are plans underway to convert millions of hectares of tropical rainforests into additional palm oil plantations, and a major driver for this is the demand for biofuels.[45]

RAPID EXPANSION OF OIL PALM PLANTATIONS

EVEN BEFORE THE pollinator weevil was introduced in Malaysia, its palm oil production had overtaken the production of all of Africa, and when Indonesia's palm oil production increased to match Malaysia's in 1987, world production virtually exploded. In 2012 more than 85 percent of the global annual palm oil production of 50.2 million metric tons (according to FAO statistics) came from only two countries: Indonesia and Malaysia (see figure 4.6). According to a study by the International Institute for Sustainable Development, production is expected to grow by another 40 percent by 2020.[46] Establishing oil palm plantations is now a government policy in many tropical countries, including some of the poorest.[47] Since the oil palm originates from the tropical rainforest and grows best in lowland rainforest climates, the consequences of such policies are not difficult to forecast.

FIG. 4.6. World production of palm oil 1980–2012

Indonesia and Malaysia together accounted for more than 85 percent of world production in 2012. The Southeast Asian oil palm growing area increased twelvefold in this period. Data from FAOSTAT.

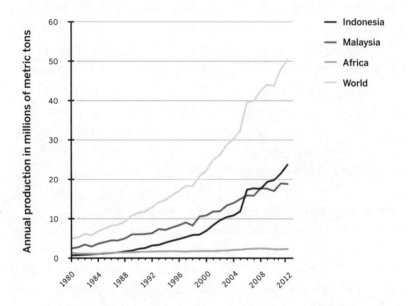

The area of Southeast Asia occupied by palm oil plantations increased twelvefold from 1980 to 2012. And there are no signs of a slowing trend; since 2000 the area planted with oil palms in Southeast Asia continued to grow almost linearly. In Indonesia it more than tripled between 2000 and 2012, reaching a total area of 6.5 million hectares—an area the size of Sri Lanka. Indonesia overtook Malaysia in planted area as well as production in 2005, but Malaysia also continued to expand its oil palm plantation area to reach 4.3 million hectares by 2012.[48] Both countries plan to extend their oil palm cultivation through 2020, although Malaysia is reaching the limits of further oil palm expansion.[49]

Oil palm plantations are primarily limited to lowland areas. In these two main producer countries most palm oil development to date has been at the expense of lowland rainforests, including peat swamp forests. Until recently the Indonesian government promoted forest conversion for oil palm plantations by granting long-term leases to influential individuals and families. Selling the timber from cleared land provides starting capital for the oil palm plantation and is a very profitable business. The Indonesian Sinar Mas Group, owned by the powerful Widjaja family, benefits from both the ownership of one of the largest paper producers in the world, Asia Pulp and Paper, as well as Golden Agri-Resources, a holding company for several palm oil producers.[50]

LOOMING USE OF PALM OIL FOR BIOFUELS

INDONESIA HAS A ready use for its palm oil as biodiesel for its domestic transport sector. But biodiesel from converted rainforest areas also seems poised to become an internationally traded commodity. The main boost for palm oil as biofuel happened in 2005–06 when petroleum prices increased beyond the price of crude palm oil. However, since then palm oil has again traded at higher prices than those of petroleum. The demand for palm oil as biofuel is therefore more a function of government policy than price.[51]

Biodiesel production is expected to increase by 50 percent to about 42 billion liters in 2020, according to the Organisation for Economic Co-operation and Development (OECD). If a considerable part of this increase were to come from palm oil, this could further accelerate forest loss in the producer countries. The International Institute for Sustainable Development found that palm oil made up a larger proportion of the subsidized European Union biofuels industry than previously estimated. Between 2006 and 2012, the use of palm oil increased from 0.4 million to 1.9 million metric tons per year. The institute estimates that government support to biodiesel in the European Union was in the range of EUR 4.6–5.6 billion (USD 6.4–7.8 billion) in 2011.[52]

THE FUTURE OF OIL PALM GROWING

THE GLOBAL DEMAND for palm oil has been increasing almost exponentially over the past thirty years and does not show any weakening. As palm oil is relatively cheap and has many uses, including as biofuel, the demand will continue to increase, mostly driven by the increasingly affluent and urbanized populations in emerging economies. If palm oil production increases by 40 percent by 2020 as expected, it is evident that more and more land will be covered by oil palm plantations, even if improvements in productivity counteract the expansion to a certain degree. The expansion will continue to be at the expense of lowland rainforests on all tropical continents. Economic growth and food security clearly outweigh biodiversity conservation in national policies, and in many areas oil palm cultivation is a valuable smallholder crop, which further increases the social pressure for conversion of natural forests.

As suitable land becomes scarce in Southeast Asia and a two-year moratorium on new oil palm concessions in intact forest has been introduced in Indonesia in 2011, African rainforest areas are being targeted for oil palm expansion. Indonesian and Malaysian companies are extending their palm oil production areas into Africa and South America. Among these companies is the Indonesian-held Golden Veroleum, a subsidiary of Golden Agri-Resources, which is part of the powerful Sinar Mas Group belonging to the Widjaja family, as well as the Malaysian palm oil giant Sime Darby and the Malaysian State Plantation Agency. Wilmar International and Olam from Singapore are in the process of establishing oil palm plantations in West and Central Africa. European corporate groups are also active across Africa, expanding industrial oil palm plantations in the twenty-three African countries with suitably moist climates for the oil palm. Many national and international development agencies are actively involved in the promotion of palm oil plantations in these countries. The World Rainforest Movement estimated that about 1.2 million hectares were under industrial plantation in 2010 in addition to the plantations of smallholders who plant oil palms as part of their traditional agroforestry systems, a combination of food crops and trees.[53]

The patterns of the oil palm industry development in Southeast Asia of the past decades suggest that in countries with weak governance and uncertain land tenure rules, oil palm expansion is more likely to happen at the cost of natural forests and to the detriment of smallholder agroforestry systems.[54] We need more responsible corporate behavior from the large palm oil producers and much stricter standards for oil palm plantations. Palm oil development should preserve biodiversity by protecting forest areas with high conservation value, maintaining buffer areas along watercourses, and similar measures. Such standards have been developed by the Roundtable on Sustainable Palm Oil (see box 4.3), and ideally they will become the norm for all future palm oil production. Other measures have been suggested as well: more emphasis on yield optimization to diminish the pressure for further expansion; promoting good governance to protect smallholder tenure systems and encourage conservation; and supporting smallholder organizations that can stand up against large-scale commercial interests.[55]

BOX 4.3   The RSPO

In 2001, increasing concerns over the rapid loss of rainforests for the
conversion into oil palm plantations led the WWF to explore the possibilities for creating a group to guide sustainable palm oil production. The
result was the Roundtable on Sustainable Palm Oil (RSPO), an informal
cooperation among Aarhus United UK Ltd; Switzerland's largest retailer,
Migros; the Malaysian Palm Oil Association; and Unilever, together with
the WWF in 2002. The first roundtable took place in Kuala Lumpur in
2003 and was attended by 200 participants from sixteen countries. The
following year, the RSPO was formally established under the Swiss Civil
Code with a governance structure that ensures fair representation of
all stakeholders throughout the entire supply chain.

The RSPO standard requires environmental impact assessments
and provisions to protect high conservation value forests (see box 7.4)
and forests on peatland. Oil palm plantations do not necessarily have
to come at the expense of tropical rainforests. Indonesia has a lot of
Imperata cylindrica (cogon) grassland and unused cleared land that
can be planted without clearing more forest or infringing on smallholder land.

The RSPO came under harsh criticism from a number of environmental groups. They accused the RSPO of greenwashing palm oil
producers that did not bring a lasting improvement to people's lives
and criticized the failure to adopt stricter emission standards by ruling
out all conversion of peatland. Some of the largest buyers of palm oil,
such as Unilever, have since gone beyond the RSPO standard and sanc-
tioned suppliers that used doubtful practices. The RSPO has made an
important contribution to more sustainable palm oil production, but
producer countries should no longer rely on voluntary standards alone;
they should incorporate non-deforestation criteria in their national
policies.

World meat consumption devastates South American rainforests

SO FAR, RAINFOREST conversion for soybean cultivation and cattle pastures is essentially a Latin American phenomenon—primarily Brazilian. Soy and beef production have not become major deforestation drivers in Southeast Asia or Africa. Three main factors may have prevented this development in the old world:

1) Neither soybeans nor cattle are well adapted to tropical rainforest climates and low-nutrient soils, so that their introduction requires specialized agricultural practices and technology.
2) Such reforms are more difficult to introduce in culturally diverse smallholder communities, which are more prevalent in old world rainforest areas.
3) Land tenure systems in Asia and Africa hold most forest land as state land, often under the control of smallholder communities. It is more difficult for large-scale enterprises to acquire land there than it is in South America.

With the enormous demand for food and the pressure of multinational companies, these circumstances may change. The turbulent history of soy and beef production in Amazonia may offer some important lessons.

SOYBEANS FEED ANIMALS AROUND THE WORLD

SOY EXPANSION IN South America started in the early 1990s with the rapid increase of meat consumption and the demand for protein-rich feed for chicken, pigs, and cattle. Fish meal, an important cattle feed, was no longer available after the collapse of the Peruvian anchovy fisheries.[56] Soybean cultivation started in the wooded grasslands of the cerrado south of the Amazon rainforest area, mainly in the southern parts of the Brazilian state of Mato Grosso. At first, the direct impact of soy growing on the Amazon rainforest was relatively modest, but even so, soy cultivation raised land prices and displaced cattle ranching deeper into the Amazon.[57]

FIG. 4.7. Soybean price in relation to deforestation

Brazilian soybean price per metric ton 1992–2011 (in 2004–2006 USD) in relation to annual deforestation in the Brazilian Amazon (in 10,000 hectares). Soybean prices from FAOSTAT. Deforestation data from the Brazilian National Institute for Space Research.

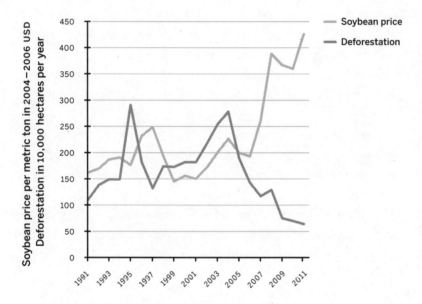

Toward the end of the last century, new varieties of soybean were introduced that tolerated the humid tropical climate of the Amazon Basin better, and soy cultivation started pushing farther north. Between 1990 and 2005 the area of soy cultivation doubled. About half the huge quantities of soy harvested around 2005 were used domestically, and livestock production grew in parallel, albeit not as fast. The other half of the production was exported, making Brazil the largest soy exporter in the world.[58] In the first years of the new millennium, deforestation in the Brazilian Amazon seemed increasingly a function of the price of soy.[59] Rising soy and beef prices became a strong incentive to clear rainforest in the southern and eastern Amazonian states, causing what is now known as the "arc of deforestation." This development continued rapidly, causing widespread concern in the international community, until the tide started to change in 2005–06 (see figure 4.7).

FBOMS, a working group of nongovernmental organizations and other Brazilian institutions, analyzed illegal land use change in the arc of deforestation, mainly in Mato Grosso between 2001 and 2003.[60] They found that soy cultivation tended to displace cattle ranching farther north, most often at the expense of closed forests. The transportation corridor for soy from Porto Velho, Rondônia, on the Rio Madeira to Itacoatiara, a deepwater port on the Amazon River, and along the BR-163 highway from Cuiabá in the cerrado area of Mato Grosso to Santarém, which has been paved in the meantime, also became susceptible to deforestation for soy plantations (see plates 8 and 9). A clear correlation between soy plantation and the rate of Amazonian deforestation was established, and FBOMS's ten-year forecast was that the area planted with soy could more than triple by 2014, depending on the market situation. It did not come to this.

THE SOY TIDE TURNS

CLEARING RAINFORESTS FOR chicken, pig, and cattle feed was obviously neither environmentally nor socially sustainable. But arguments based on sustainability alone would have been unlikely to make a real difference. Easily accessible satellite imagery made the public a witness to a problem of global dimensions that increasingly tarnished President Lula da Silva's government. Companies' concern for their public reputation became the key driver for change as people, particularly in the European Union, realized that soybean plantations and cattle pastures in Amazonia were driving deforestation.[61]

When a Greenpeace campaign in 2006 targeted international food and agricultural commodity companies, accusing them of deforestation, slave labor, and illegal infringement in indigenous territories, more than sustainability was at stake for these companies. Cargill, one of the largest multinationals involved in soybean expansion and processing in the states of Mato Grosso and Pará, was the main supplier of soy-fed chicken to the fast-food chain McDonald's, through its subsidiary Sun Valley. Both multinationals, Cargill and McDonald's, saw their reputations at serious risk and reacted swiftly to the well-documented Greenpeace report (*Eating Up the Amazon*),[62] promising deforestation-free products in future. The largest soy producers had no choice but to agree to a moratorium on buying soybeans from recently deforested land. The moratorium was carried by the Brazilian Association of Vegetable Oil Industries (ABIOVE) and the National Association of Cereal Exporters (ANEC) and has since been extended annually.[63] This moratorium has worked: by 2012 only minimal areas were deforested for soy cultivation, while soybean production continued to grow.

DECOUPLING DEFORESTATION FROM THE SOY PRICE

THE CASE OF soybean cultivation in Amazonia is a particularly interesting example that may help us understand the dynamics of deforestation. From 1996 to 2005 the Brazilian states in the arc of deforestation—Mato Grosso, Rondônia, and Pará—accounted for 85 percent of deforestation in the Amazônia Legal region, with an average of 1.66 million hectares of forest loss per year.[64] After 2005, however, deforestation started dropping dramatically (see figure 4.7).

The decline in deforestation was particularly remarkable in the most important soybean cultivation state of Mato Grosso, where soy production continued to increase nevertheless. A detailed analysis of land use change in Mato Grosso[65] showed that soy production became decoupled from deforestation after 2005. Two main factors contributed to this decoupling:

1) Soy production expanded on previously cleared areas, mainly pastures; to a lesser degree farmers achieve higher yields per area.

2) Policies were implemented to restrict credits for deforestation, and monitoring and enforcement were improved through the Amazon Deforestation Monitoring Project (PRODES), which created a powerful disincentive for deforestation. The public campaign and the resulting voluntary moratorium on soy from recently cleared areas triggered these developments.

An important question about the decrease of deforestation after 2005 remained, however: could there have been "leakage," that is, displacement of soybean plantations to other areas? The analysis cited above[66] showed that the decline of deforestation in Mato Grosso after 2005 did not cause a soy expansion in the cerrado areas of the southern parts of Mato Grosso, and that in the adjacent Amazonian states of Rondônia and Pará, deforestation declined as well. The lack of obvious leakage into other areas may have been due to soy expansion into previously cleared low-productivity pastures.

Beef from rainforest soils: The most inefficient land use ever

IN THE SAME tropical rainforest areas affected by soybean cultivation, in the states of Rondônia, Mato Grosso, and Pará, another industry has contributed even more strongly to the Brazilian arc of deforestation: cattle. Though cattle ranching is not independent from soy expansion and has followed the same course of export-driven dynamics, pushing northward into the Amazon, it shows a number of different characteristics. Cattle ranching needs a lot of space, and extensive former rainforest areas are therefore used for cattle production. It has low investment per hectare and shows a pattern of frequent land abandonment.[67] Such wasteful expansion for very low returns was only possible because land in the north was cheap, and illegal pasture clearing was common until some years ago. The very low stocking density of about one cow per hectare combined with the slow growth rates of the animals results in minimal productivity. Even in the European Union, where the productivity per hectare is higher, beef production is very inefficient: per calorie produced, beef production requires twenty times as much land as pork, eighteen times as much as poultry, and seventy-nine times as much as cereals production.[68]

In addition to the extensive land requirements of cattle ranching, another major environmental impact has to be considered: cattle pastures are cleared with heavy machinery, and hardly any of the timber is used. The cut trees are usually left to dry, and the desiccated vegetation is burned during the dry season. Cattle pastures are also burned in subsequent years, as this triggers the growth of new grass shoots at the beginning of the drier period of the year.[69]

Even without taking the emissions from land clearing into account, beef production releases a high volume of greenhouse gases per calorie. In the European Union, beef production emits seven times as much greenhouse gas per calorie as pork, nine times as much as poultry, and seventy times as much as cereals. This is mainly due to the large amount of methane emitted by ruminant feed digestion.[70] Considering the large land requirement of cattle pastures, the practice of clearing land by fire, and the direct greenhouse gas emissions by cattle, it is not difficult to imagine that producing beef in cleared Amazon forest areas has a carbon footprint far more than ten times that of producing the same amount of pork or poultry, let alone cereals or vegetables.

DRIVERS OF CATTLE PASTURE EXPANSION

BETWEEN 1990 AND 2002 the number of cattle in Amazonia more than doubled. The cattle industry, as a consequence, was responsible for more than two-thirds of the annual deforestation in the Brazilian Amazon. Two underlying factors have contributed to this rapid expansion:[71]

1) From 1998 to 2002 the Brazilian real fell to a third of its value against the dollar, which doubled the price received for selling beef to foreign markets and created an incentive to clear more land for cattle pastures. At the same time the price of Brazilian beef in dollars dropped, which made exports more competitive.

2) Previously, foot-and-mouth disease in most of Brazil had prevented beef exports to international markets. By 2003, eradication programs had ensured that 85 percent of the country's cattle lived in areas that had been certified free of foot-and-mouth disease. This boosted beef exports to Europe, Russia, and the Middle East. Brazilian beef exports increased sevenfold in less than a decade, and Brazil became the world's largest beef exporter in 2004.

Around 2004 and 2005 a number of scientific observers estimated that if current deforestation trends continued, by 2050 the closed canopy forest of the Amazon Basin would be reduced to 320 million hectares—about 53 percent of its original area. In this scenario a quarter of the 382 mammalian species would lose more than 40 percent of their habitat, and thirty-five primate species would lose between 60 and 100 percent of their habitat. Deforestation would cause the emission of 32 ± 8 billion metric tons of carbon, which corresponded in 2005 to about four years of carbon emissions worldwide.[72] The projection was based on the deforestation rates caused by soy and cattle expansion as they were actually observed during the peak years of deforestation in the Brazilian Amazon. Linear growth projections over decades, of course, are not predictions—they only indicate the rate of change. Still, hardly anybody could have anticipated that progress would come as quickly as it did in the years after the turning point in 2004 (see figure 4.8).

FIG. 4.8. Brazilian soybean and cattle production 1990–2012

Production of cattle meat and soybeans compared to deforestation. Production is indexed to 1990 values of 19.9 million metric tons of soybeans, and 4.12 million metric tons of cattle meat. Annual deforestation in the Brazilian Amazon in 1990 was 1.37 million hectares. Soybean and cattle meat production data are from FAOSTAT. Deforestation data are from the Brazilian National Institute for Space Research.

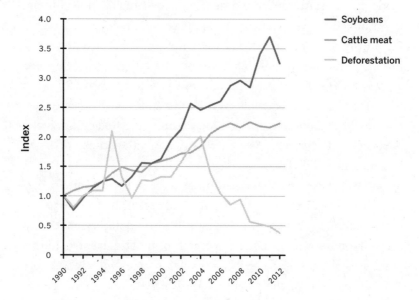

THE BEEF MORATORIUM

AS HAPPENED WITH the public outcry against soy expansion in 2006, the wasteful destruction of forest for cattle pastures came under harsh criticism just three years later, in 2009. Two organizations, the Brazilian nongovernmental organization Amigos da Terra—Amazônia Brasileira and Greenpeace, played a key role. They divulged the practices of the meat industry and accused the largest slaughterhouses and meat processors Bertin SA, JBS (which acquired Bertin briefly thereafter), and Marfrig SA of involvement in illegal Amazonian deforestation.[73] They also accused the Brazilian National

Development Bank of providing the funding for expanding pastures into forest areas. The International Finance Corporation (IFC), the private lending branch of the World Bank, which had approved a USD 90 million loan to Bertin SA for cattle pasture expansion and processing, quickly canceled the loan in 2009.[74] The IFC had approved this loan against the advice of its own Independent Evaluation Group!

As was the case with the companies that committed to a soy moratorium, the large slaughterhouses were forced to henceforth buy only from ranches that could prove they had no history of deforestation. The federal prosecutor's offices in Pará and Mato Grosso monitored deforestation by requiring ranches in their states to provide GPS coordinates of the boundaries of their properties. Brazil's largest supermarket chains, Carrefour, Walmart, and Pão de Açúcar, were also warned they could face prosecution if they continued to buy from slaughterhouses that could not bring proof that their products were from legal sources.

An important international link was established through the leather industry, which supplied such international brands as Adidas, Puma, Nike, BMW, and many other car companies, as well as other large European and United States retailers. Though leather was only a by-product of the beef industry, the Leather Working Group,[75] established a few years earlier to set environmental standards for tanners, was called upon to address the problem of deforestation and to set a zero-deforestation standard. Consumer pressure expressed and amplified through the campaigns of nongovernmental organizations combined with the intervention of the Brazilian authorities led to an impressive change in the business practices of the major corporate players.

BEEF CONSUMPTION STILL A KEY DRIVER

IT MAY BE too early to celebrate success. Other South American countries that share the Amazon rainforests with Brazil also show signs of cattle pasture expansion into forest areas. In Colombia and Ecuador deforestation for pastures has been mainly caused by small-scale ranches until recently, and their export market is comparatively small (see Cláudio Maretti's Specialist's View on p. 204). However, in 2013 the IFC acquired a minority stake in the Brazilian company Minerva SA and provided a loan of USD 60 million to build slaughtering and processing capacity and to expand into other South American countries.

This IFC involvement may change the circumstances as it also covers expansion into Colombia, which does not have a land registry and monitoring system. After the cancelation of the IFC cattle production loan to JBS-Bertin in 2009, this came as a surprise and immediately triggered protests from the Sierra Club and other nongovernmental organizations.[76] The IFC's previous involvement in cattle expansion in Amazonia does not seem to have led them to reconsider their policies. Given the far-reaching consequences of beef production in rainforest areas, it is difficult to comprehend how such lending practices could ever be reconciled with the declared objectives of the IFC.

The history of soy and beef in Amazonia holds lessons for the world

THE DEFORESTATION IN the Brazilian Amazon caused by commercial soybean cultivation and beef cattle production has declined drastically since 2004, positively influencing the deforestation statistics not only of Brazil, but of the whole region. We can draw some lessons from this experience that could well be relevant for other tropical rainforest countries:

1) The underlying causes of deforestation were commercial. The commodities produced were at least partly destined for highly sensitive international markets, which were exposed to consumer attitudes.

Consumer pressure affected the corporate reputation of producers and ultimately their profitability.

2) Government policies regulating deforestation were introduced; these included restricting loans to projects that would cause deforestation, and prosecuting violators. Customers of illegally produced commodities were also threatened with prosecution.

3) Technological advances in remote sensing made it possible to support enforcement through surveillance with the PRODES monitoring program.

4) Increasing productivity per area by using better plant varieties, soil and pasture improvements, higher stocking rates, and more feed lots allowed a more efficient use of already cleared land.

The real lesson to be drawn from the Brazilian success story is that it has not resulted in lower production. On the contrary: Brazilian soybean and beef production continued to increase after 2004 when Brazilian deforestation rates started to decrease steadily (see figure 4.8).

The battle against deforestation may not be won yet. Deforestation increased slightly, by 28 percent from 2012 to 2013. Though 2013's deforestation rate was still the second-lowest since 1988, this increase evoked fears that this could be a resurgence of deforestation in the Brazilian Amazon. Some environmentalists and media attributed this to revisions in the Brazilian Forest Code in 2012, which permit a reduction in the protected forest on farmland and introduce an amnesty for land cleared before 2008, although this seems an unlikely hypothesis.[77]

The jury is still out on whether the decline in deforestation will be sustained, and it should be noted that while deforestation rates in the Brazilian Amazon have been drastically reduced, land clearing in the wooded cerrado areas may also have taken a heavy toll on areas of rich biodiversity.

RAPID WORLD URBANIZATION: GOOD OR BAD FOR THE RAINFORESTS?

FROM THE 1960s until the early 1990s the major forces behind deforestation were slash-and-burn farming and state policies aimed at economic development. Those commonly drove land use change through credits, tax cuts, and incentives for commercial crops or pastures as well as infrastructure development, such as roads. But from the early 1990s onward, state influence started waning and became increasingly replaced by enterprise-driven development.[78]

Concurrently with the shift from state actors to enterprise-driven land use in tropical forest countries, another pattern has become increasingly evident: tropical rainforest loss correlates with urbanization and agricultural exports.[79] Those countries with the highest forest loss are also those with high urban growth and the most intensive agricultural trade. All tropical rainforest countries, moreover, are urbanizing rapidly. According to the 2011 Revision of the United Nations' "World Urbanization Prospects" report, by 2050 the urban population will reach 64 percent in Asia, 58 percent in Africa, and 87 percent in Latin America. Over 70 percent of the world population is expected to live in urban areas by 2050.

Some authors have expressed hopes that slowing population growth and rapid urbanization would take demographic pressure off tropical forests and slow deforestation. They anticipated that with fewer people living in rural areas, forest regrowth could accelerate, thus avoiding mass extinctions of tropical forest species in the coming decades.[80] The link between rural population densities and deforestation has been a commonly held belief in the past. This may still apply to some of the poorest rainforest countries inhabited by immigrant subsistence farmers, as for example in Central Africa, but in a globalized economy where trade in agricultural commodities is dominated by international corporations and financial flows, forest loss does not decline as populations move from rural to urban areas. In fact, the inverse is true.

CARNIVOROUS URBAN CENTERS

URBAN POPULATIONS TEND to change their food habits. With rising incomes, urban populations consume less staple food and more processed food and animal protein—meat and dairy products—than rural populations. The production of meat and dairy products for these urban markets is commercial and more land-intensive. It tends to be located in other geographic regions, and products are traded internationally.

Meat production is rising steeply (see figure 4.9), particularly pork and poultry. Even though these farm animals have a much lower land requirement than cattle, meat statistics hide the amount of land used to produce grain for animal feed. The increase in world meat production is partly driven by population growth, but it is more closely related to the demand from the urban populations of wealthier developing countries.

On the other hand, the production of beef shows signs of leveling off, mainly because of the lower demand in industrialized countries. The per capita beef consumption in the United States has decreased by more than 40 percent since the 1970s, and beef production in the most important production countries is stagnating (see figure 4.9). Meat production has a variety of negative environmental impacts, not least greenhouse gas emissions, but the example of Brazil has shown that meat production does not have to come at the expense of additional tropical rainforest areas. With the right kind of policies it would be possible to cover the rising world meat demand of the next decades—a demand that is likely to start decreasing in the second half of the century—without further rainforest destruction.

FIG. 4.9. World meat production and beef production in the five most important producer countries

While world meat production is rising steeply mainly because of higher income levels in Asia and Latin America, the proportion of beef is in decline, and production in the five largest producer countries seems to be leveling off. Data from FAOSTAT.

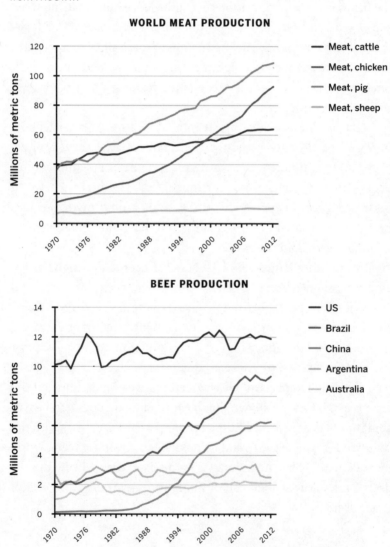

A population shift to urban areas tends to increase food waste. An estimated 30 percent of food is wasted currently.[81] In industrial nations a large proportion of food produced for human consumption is discarded, often from individual households and restaurants, while in developing nations food waste occurs primarily in the after-harvest storage, transportation, and processing.[82] With increasing urbanization, food waste could rise further and affect future food requirements, estimated by the FAO to increase by 60 percent by 2050.

The ecological footprint of urban centers is substantially larger than the sum of the individual rural parts. Global markets for livestock feed (such as soybeans), beef, and vegetable oils (mainly palm oil) connect the urban centers with large plantations established at the expense of tropical rainforests. These links, now often referred to as "teleconnections," also explain the observation mentioned above: that it is not state actors anymore that drive land use change in many countries, but rather large commercial enterprises and international financial flows.

JÜRGEN BLASER

Sustainable Management of Tropical Forests: From the Ideal to the Possible

SUSTAINABILITY AND FOREST MANAGEMENT

For more than two decades sustainability has been a global political priority, and sustainable forest management has become an essential tool. Sustainable use is use of biological and social systems that does not impair their availability to future generations. This means that ecological conditions do not deteriorate, that productive capacity will support future generations, and that the capacity of the social system to resolve conflicts and create institutions will be no worse than today.[1] Sustainable forest management became a new paradigm for forest management as countries recognized how many goods and services come from the forests and—at least at the level of political intention—started moving away from simply exploiting forest resources and advocating single-use management. However, sustainable forest management is not a static approach, but a constantly evolving and adaptive process with an ever-expanding scientific basis. It is about balancing the different uses of the forest while ensuring continued ecological functioning and provision of ecosystem services.

WHAT CONSTITUTES SUSTAINABLE FOREST MANAGEMENT?

A strategy for sustainable forest management must reflect a variety of objectives:
• continuously satisfy the needs for timber, fiber, and nontimber forest products ("forest goods");
• ensure conservation of soil, freshwater flow, and carbon stocks;
• sustain the resilience of forest ecosystems and renewal capacity of all species;
• conserve biological diversity;
• support the food security and livelihood needs of local communities; and
• provide a fair sharing of the benefits from forest uses, including financial, cultural, spiritual, and recreation values.

Hence, sustainable forest management considers the forest in both time and space and represents a balance between conservation and producing forest goods and services for humans. It must operate within the capacity of the forest to recover and maintain its functions.

Recent developments in forest policy, particularly REDD+,[2] present new opportunities to advance and implement sustainable forest management, particularly in the tropics. At the same time, the concept has been criticized. Nongovernmental organizations have brought forward concerns about logging in primary forests and about the establishment and management of forest plantations.

Spurred by the increasing acceptance of REDD+ in climate change negotiations and by the potential role of sustainable forest management in the development of a green economy, there is a need to increase the common understanding of the sustainable forest management concept. There is also a need to agree on the values that need to be traded off when using forest goods and services so that forests can contribute effectively to the sustainable development goals of the global community. The World Commission on Forests and Sustainable Development[3] stated that sustainable forest management "must be a flexible concept that accepts changes in the mix of goods and services produced or preserved over long periods of time and according to changing values signaled by various stakeholder groups," and that sustainable forest management "should be viewed as a *process* that can be constantly adapted according to changing values, resources, institutions and technologies."

Sustainable forest management in tropical primary forests

The thirty-three member countries of the International Tropical Timber Organization[4] contain about 90 percent of all natural tropical rainforests. They report an area of 761 million hectares of forests classified as "permanent forest estates." Permanent forest estates are land, whether public or private, that is secured by laws and kept under permanent forest cover.[5] About 403 million hectares of permanent forest estates are classified as production forests, and 358 million hectares as protection and conservation forests.[6]

Out of these forests about 500 million hectares are primary forests, which means that they have been unaffected by logging and other disturbances and that their natural structure, functions, and dynamics have not undergone any changes that exceed the elastic capacity of the ecosystem.[7] Around 165 million hectares are subject to long-term timber concessions.[8] At least half these concessions are operating in primary forests in the first logging rotation (the first cut). About 40 percent of concessions are operating in the second cut, and 10 percent of concessions have already undergone several logging cycles over the past decades; they no longer fulfill the definition of "primary or close to primary forests." While about 130 million hectares of these concessions were managed according to approved management plans in 2014,[9] only about 26 million hectares are certified, meaning they are "well managed" according to defined certification standards.

The management of tropical rainforests is complex. Primary forests regenerate in small patches (gaps) and are ecologically multifaceted. Thus the first logging activities in such complex ecosystems are decisive for the destiny of these forests. Will they remain as close-to-primary forest or will they become ecologically and economically degraded? Tropical primary forests exhibit particular features that need to be taken into account when logging them:

- Tropical forests feature emergent trees (trees that extend above the main forest canopy). They often have diameters exceeding 150 centimeters, and they have grown over centuries. Many of these trees have high commercial value and are the preferred target of logging in primary forests, particularly in the first cut. However, as these trees have grown over long periods of time, they should not be part of consecutive cutting cycles in a managed "primary" forest.

- Sites and forest types show great variety, with different structures and composition. Hundreds of tree species can occur in a forest type, and even within the same forest type the mixture and structure of the forest stand can change over a small area. This makes forest management complex and challenging.

- Most of the commercially interesting species occur in small numbers throughout the area. Of the commercially interesting species, only a few individuals will emerge above the canopy layers. However, in all tropical forest types there are tree species that occur in more regular patterns, often with more individuals per species in all sizes. These species are of particular interest in managed forests, although they are often not the preferred species from a commercial viewpoint.

- Not very many tree species in tropical rainforests produce marketable wood (5–20 percent), with the exception of the Dipterocarp forests in Southeast Asia. Many of the commercial species are characterized by irregular diameter ranges, which means they occur as dominant top canopy trees, while they are often rare as medium-sized and small trees and as seedlings in regeneration. They are generally light-demanding species that have installed themselves in the few bigger gaps or during an earlier secondary forest cycle. Thus, although there is natural regeneration in primary forests, the abundance of economically valuable species is disproportionately low.

Concerns about managing intact primary forests particularly focus on the long-term effects of logging. Indeed, with the exception of some particular forest types—certain Dipterocarp forests in Southeast Asia and forests dominated by few species, such as peat and swamp forests, mangroves, and tropical conifer forests[10]—the majority of humid tropical primary forests do not easily fulfill the criteria for sustainable management. They would need to be gradually transformed to homogenize their composition and structure. However, recent research has shown that tropical managed forests can retain high biodiversity, often as high as undisturbed forests.[11] Accordingly, in largely intact forest landscapes where there is currently little deforestation and degrada-tion, the conservation and management of existing forests is critical both to prevent future greenhouse gas emissions through loss of car-bon stocks and to conserve biodiversity.

Nonetheless, the experience over the past thirty years has shown that even the application of good forest management practices in intact primary forests leads to biodiversity loss and requires some trade-offs with regard to carbon emissions. This is mostly because of the increased risk of deforestation and overhunting, both of which are closely associated with the proximity to access roads.[12] The prob-lems associated with logging primary forests therefore do not directly result from sustainable forest management itself, but rather from governance problems and poverty-driven slash-and-burn agricul-ture. For example, the likelihood of deforestation in logged forests in the Brazilian Amazon was found to be up to four times greater than for nonlogged forests.[13] It is believed that in Southeast Asia, building logging roads to harvest valuable Dipterocarp trees in lowland forests has led to deforestation in sparsely populated protected areas like Kalimantan.[14]

Given the practical considerations of sustainable forest manage-ment in natural tropical forests, it may be more practical to make forest management decisions that will contribute incrementally to more sustainable forest management. In other words, avoiding *unsus-tainable* forest management may be a better goal than trying to attain a vague ideal of perfect sustainable forest management. In many developing countries the annual harvest (of timber, fuelwood, and non-timber forest products) is much higher than is sustainable because of low environmental standards and lack of enforcement of regulations. Therefore, most tropical forests need continuous improvement in how they are governed and managed. Management objectives can be based on best practices, but even if they are not perfect they can still guide forest management decisions.[15]

For those who are more concerned with the practical application of forest management and less with the theoretical details of sustainable forest management, such an approach is much more sensible. It could be referred to as *practical sustainable forest management,* defined as "the best available practices, based on current scientific and traditional knowledge, which allow multiple objectives and needs to be met, without degrading the forest resource."[16] Taking such an approach to sustainable forest management in tropical natural forests is important, considering that most forests outside effectively protected areas have been, or are at risk of being, selectively logged. It is essential to maximize the conservation values of partially harvested areas.[17] Tropical natural forests, if managed through silvicultural stand improvement methods, can retain substantial biodiversity values, timber, and carbon stocks. These methods can be viewed as the second-best option after full forest preservation, and they deserve much more attention from conservationists, researchers, and policy makers.

Sustainable forest management in degraded and secondary forest

Degraded and secondary forests are now the predominant forest types in many tropical countries. Degraded forests are skimmed off primary forests where timber, fuelwood, and other forest products have gradually been depleted. Secondary forests contain various stages of succession and are less heterogeneous within and between sites, at least during the early pioneer stages.[18] They are also less diverse than primary forests. The dominant species in the early secondary stages are short-lived pioneer trees that are extremely light-demanding. Over time, secondary forests become more diverse and shade-providing species can install themselves as long as their seed stock and seed dispersers are still present. The biomass of such forests can reach that of primary forests in the course of a hundred years or more, depending on the site conditions. Under good site conditions, secondary forests have a high capacity to sequester carbon dioxide and can become important carbon sinks. Under unfavorable conditions, however, the intensity, frequency, and scale of the disturbance may push a degraded forest over an ecological threshold so that recovery is slow or impossible and the site continues to degrade.

Degradation is often considered to be a precursor to complete deforestation and is considered to be part of the same process. While it is true that in some cases degradation is followed by deforestation, this tends to be the *exception* rather than the rule. For example, it has been observed in the Brazilian Amazon and the Congo Basin that commercial logging (as the degradation driver) may be followed by agricultural clearing as migrant farmers move in along the logging roads. In many other places degradation is not caused by commercial logging but by extraction of various forest products, often for subsistence or for local markets (timber, firewood, charcoal, bushmeat, and fodder), or by shifting agriculture by forest dwellers and indigenous communities. In such areas degradation very rarely leads to deforestation, but rather to a gradual and substantial loss of carbon stocks, along with the loss of other forest goods and services.

Because of the instability of production and the often low value of timber, degraded and secondary forests often do not receive the benefit of sustainable forest management, at least not in the short term. With REDD+ and the potential to restore carbon stocks in degraded and secondary forests, this might change in the near future. Secondary forests should be given more attention in forest management as they can produce, in a relatively short time, light-demanding hardwood species that have commercial value. A number of light-demanding secondary species such as okoumé, ayous, limba, and fraké in Africa; cordia, cedrela, and jacaranda in tropical America; and albizia, kadamba, and others in Southeast Asia are also capable of supplying valuable timber in quantity and quality, not unlike the currently much-used teak.

Managing degraded and secondary forests is often associated with the term "forest restoration." Forest restoration is a term increasingly being applied in the forest policies of many tropical countries. In degraded forests, the number of economically and socially desired species present in a degradation stage is seldom adequate and requires additional interventions. Stimulating natural regeneration and, in certain cases, enrichment planting[19] are appropriate treatments for forest restoration. The key question is how much tending is required and how the cost–benefit ratios compare to those of other land uses such as forest plantations or agricultural crops. Overall, managing degraded and secondary forests is becoming more and more important in many tropical countries. Encouraged by REDD+ and other initiatives, community-based forest management and farm forestry are focusing on such forest areas. Silvicultural treatments can be used to restore lost carbon pools in degraded and secondary forests. Forest restoration merits political interest, particularly in countries that have lost most of their primary forests. These countries need their remaining degraded forests to protect watersheds and landscapes that are essential for the production of food and crops and the welfare of people living downstream.

Sustainable forest management and the role of planted forests

Planted forests are established by seeding or planting native or introduced tree species, either on formally forested land (reforestation) or on land that was not previously forested (afforestation). Most of these forests have been planted for productive purposes, mainly for wood and fiber. Even though planted forests only constitute 7 percent of the world forest area, they currently cover more than half the global industrial wood demand. Compared to natural forests, they require higher investment per hectare but they also produce higher yields.

Planted forests are expected to become the major source of industrial roundwood, with an anticipated potential to produce up to 80 percent of the global demand by 2030.[20] Tropical forest plantations are generally more cost-effective as they often only serve one or two specific purposes, mainly wood production (generally industrial wood, fuelwood, fiber, and to a lesser extent timber). In spite of these favorable conditions, many of the plantation investments undertaken over the past thirty years or so have failed. They generally failed because of incorrectly assessing the ecological and economic possibilities and misjudging the social acceptance for the project. From the broader viewpoint of sustainability, forests should be planted on those sites that are not favorable for agricultural purposes. However, this will increase installation and maintenance costs and reduce profit margins.

Planted forests, along with urban forestry and agroforestry, are major elements in the solution package for a sustainable future, despite the many legitimate concerns about their ecological, social, and economic impact.[21] There is enough knowledge and experience today to avoid such negative impacts in the future. More and more degraded land will be available to undertake massive reforestation efforts. Forestry and other sectors should work toward a system to sustainably manage and harvest wood fuels as a substitute for high greenhouse-gas-emission fuels, and these new plantation schemes should consider how to increase the biodiversity and biomass of secondary vegetation. Urban forestry can play an additional role, as artificially created mixed stands of species can rapidly create new biodiversity-rich environments (though as scattered islands without real stepping stones). Urban forestry is the most neglected field in forestry today with considerable potential.

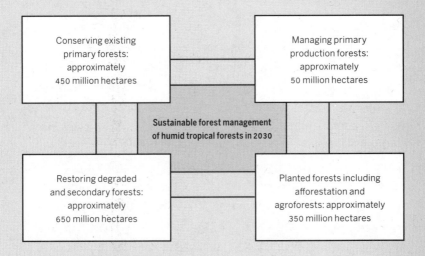

Conserving existing primary forests: approximately 450 million hectares	Managing primary production forests: approximately 50 million hectares
Sustainable forest management of humid tropical forests in 2030	
Restoring degraded and secondary forests: approximately 650 million hectares	Planted forests including afforestation and agroforests: approximately 350 million hectares

Not ideal—but practicable

Sustainable forest management in the tropics is a field in transition. Many of the timber concessions that operate in tropical primary forests will cease their operations in the coming ten years or so because of increased remoteness of operations, reduced economic viability, and a political climate that brings more primary forests under active protection. However, it can be expected that a number of operations that have been active in some areas of the tropics for several decades will become well-managed natural forests in a second or third rotation. Such operations will be specialized to deliver a small segment of high-value timber for specific purposes. They should be managed under certification schemes and be combined with biodiversity conservation, protection of carbon pools, ecotourism, and collaborative management with local communities.

The attention currently given to the management of natural tropical forests in the coming years might turn to the restoration of degraded primary forests and the management of secondary forests. These forest types are becoming the predominant forest type in nearly all humid tropical countries and merit the specific attention of policy makers. In contrast to the management of tropical primary forests for high-value timber, forest restoration can be undertaken in small forest areas, for example, at the community level with a clear focus on multiple use for timber, carbon, and other forest products and services. These forests also have the potential to complement the needs for energy and fibers that in the future will be produced more and more in forest plantations.

The implementation of sustainable forest management in tropical rainforests is far from the ideal embedded in the term "sustainability," but we need to focus on implementable and practicable solutions. It takes a diverse set of approaches to achieve sustainable forest management: conservation of large tracts of remaining primary forests; silvicultural management of other sections of forest for timber and biodiversity; restoration and management of degraded and secondary forests to protect watersheds and supply raw materials to domestic markets; and management of planted forests for wood, fiber, and energy. In combination, these approaches can substantially contribute to sustainable development. As one of the main renewable natural resources available to humanity, forests, and in particular tropical forests, can help mitigate climate change, protect soil and water, provide clean air, conserve biodiversity, help maintain the mental health of humans, and produce wood, fiber, and other products at the same time.[22]

JÜRGEN BLASER *is a professor of international forestry at the Bern University of Applied Sciences, School of Agricultural, Forest and Food Sciences.*

5

FOREST TRANSITIONS

Modest Hopes

N THE PAST century, forests in the temperate zones, particularly in Europe, have grown back after the massive onslaught driven by the energy demands of industrialization. The recovery of previously clear-cut forest areas, described as "forest transitions," gave rise to hopes that the same thing could happen in tropical forest countries. A forest transition is a net reforestation,* either by natural regrowth or forest plantation, after a period of net deforestation. The term goes back to a Scottish geographer, Alexander Mather, who published many articles on long-term changes of forests, from deforestation to reforestation, mainly in Europe.[1]

---

* The term "reforestation," when used by Mather and subsequent authors reporting on forest transitions, has a broader meaning than the FAO definition, which defines reforestation more narrowly, from a perspective of forest management, as "re-establishment of forest through planting and/ or deliberate seeding on land classified as forest." At least in this chapter the term "reforestation" is used in its broader sense (see also box 2.2).

In a number of countries, forest transitions have followed a shift to urbanization and industrialization. The case of the forest transition in Switzerland is a well-documented case, discussed by Mather[2] and many others. In 1876 Switzerland introduced a strict ban on forest clearing through one of the world's first forest laws (the *Forstpolizeigesetz*), after years of catastrophic landslides and floods, particularly in alpine areas. The Swiss forest law of 1876, with some subsequent amendments, is still in force today and has become a model for other forest legislation. It prescribes that the forest area of the country must be maintained and expanded and that any forest area cleared must be replanted in the same region.

On the one hand, the forest transition in Switzerland was rooted in sustainability considerations promoted by the rector of the Swiss Federal Institute of Technology, Elias Landolt, a forester whose convictions were influenced by the German school of sustainable forestry. He played a key role in designing and creating the new forest law.

On the other hand, the transfer to fossil fuel during the industrial revolution also affected Swiss forests, this time in a positive sense. Numerous local industries switched from wood to coal as improvements in the railway system made longer transport routes possible. The forest area of Switzerland has since increased by about 70 percent and continues to expand. Currently about 31 percent of Switzerland is covered with forest.[3] At least since the 1950s, the forest expansion is largely attributable to land abandonment by mountain agriculture and livestock pasturage in the Alps.

Following Mather's study of reforestation in a variety of European countries, U-shaped curves of forest transitions have been postulated. A period of deforestation to low levels of total forest cover would, depending upon the local circumstances and after varying time periods, be followed by a period of reforestation (see figure 5.1). Some authors tried to model this curve as a function of specific environmental and socioeconomic variables.[4]

FIG. 5.1. Forest transition theory

The forest transition theory postulates a transition from a period of deforestation in a country or region to a subsequent period of reforestation. From stage A (high forest cover and low deforestation rate) the transition passes through phases of high deforestation rates to a leveling-off of net deforestation (the forest transition), followed by stage E: net reforestation. Forest transitions are strongly influenced by national and local contexts as well as globalization effects and national policies. Adapted from Angelsen et al.[5]

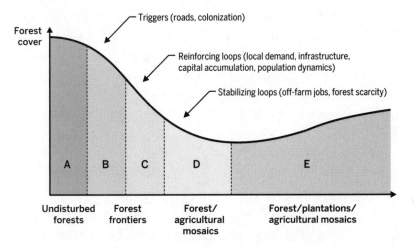

In some countries that have experienced forest transitions, the turning point of the curve happened when forest cover reached very low levels: 3 percent in Scotland, 4 percent in Denmark, and 7 percent in China. In the twentieth century, however, turning points occurred at higher levels of forest cover in New Zealand, the United States, and Costa Rica.[6] A combination of factors led to these earlier turnarounds, for example urbanization, land abandonment, and wood shortages. Perhaps the most often cited "good news" example of a forest transition is the case of China. A massive tree-planting scheme imposed by central government policy was introduced in China during the Maoist period. This scheme was reinforced with a logging ban in 1998, after the catastrophic Yangtze floods. However, the reforestation came at the cost of increased imports of illegally harvested timber, mainly from Russia and Indonesia.[7]

Tropical reforestation cases

A SMALL NUMBER of forest transitions in tropical countries have been reported since the beginning of the twenty-first century, and detailed studies of the specific circumstances that led to net refor- estation in these countries have been published. It is important to the future of tropical forests to understand what drives tropical reforestation and to judge what is different from temperate coun- tries. Ultimately we need guidelines to support policy making for future tropical forest transitions. These are a few conclusions from Central America, India, and Vietnam.

EL SALVADOR

THIS WELL-DOCUMENTED CASE of tropical reforestation took place from the early 1990s. The country had lost much of its forests even before the civil war of the 1980s. During the war, El Salvador recorded an annual deforestation rate of 2.3 percent, which increased to 2.9 percent in the 1990s. But during the same period the wood- land recovered at a rate of 5.8 percent. From the early 1990s to the early 2000s, the area with at least 30 percent tree cover increased by 22 percent, and the area with denser forest (at least 60 percent tree cover) increased by 6.5 percent.[8] El Salvador, often considered to be a devastated country, had taken a remarkable path of recovery and still harbors important biodiversity despite relatively small con- servation areas. This aspect of the biodiversity value of reforested tropical areas is an extremely important consideration for the future of global biodiversity conservation and merits closer examination.

Studying the factors that may have promoted reforestation in El Salvador, Susanna Hecht and Sassan Saatchi (2007)[9] made the interesting observation that there is no significant correlation between forest cover and population density, but there is a significant positive correlation between reforestation and remittances from Salvadoran workers abroad to their home country. Where agricultural prices are low, remittances were used by the recipient families for purposes like health, food, housing, and education rather than land clearing for agriculture. El Salvador may be a special case, given its tormented recent history, but it still emphasizes that generalizations in modeling reforestation are highly questionable.

COSTA RICA

COSTA RICA EARNED an international reputation for reforestation when it introduced a unique piece of legislation, the 1996 Forestry Law, that promotes conservation through payments for environmental services, reforestation, and conservation programs. Costa Rica experienced rapid deforestation after World War II, lasting until the 1980s.

Costa Rica's forest transition turning point was quite early for a tropical country. It occurred when about 30 percent forest cover was left, in the mid-1980s.[10] Costa Rica defaulted on its foreign loans in the early 1980s, and was subsequently submitted to International Monetary Fund structural adjustment programs and World Bank lending conditions. As a result, rural households diversified their income sources to include off-farm labor, small-scale enterprises, and the tourism industry. As was the case in El Salvador, remittances from family members working abroad (for instance in the United States) further contributed to the abandoning of economically marginal agriculture in many areas of Costa Rica. Forest regeneration on private land owned by foreigners and the activities of conservation organizations also enhanced reforestation.

Costa Rica now has about 170 parks and reserves including a large number of private wildlife reserves, covering almost 20 percent of the country. A study of Costa Rica's forest transition in the context of globalization by Kull et al. (2007)[11] showed that the forest transition is driven by a combination of factors, including real estate investment, conservation programs, socioeconomic changes in rural areas, and diversification of income, including remittances from family members abroad. These researchers concluded that Costa Rica's reforestation story needed to be told in this specific international context of markets and ideas and could not easily be compared with other forest transition cases.

Other Central American countries also show signs of forest transitions, but circumstances vary. Land cover changes in western Honduras, for example, follow complex patterns and result in changing mosaics of land use over time.[12] While some reforestation in this part of the country could be attributed to a logging ban introduced in 1987, new areas were also deforested between 1987 and 1996.

INDIA

OTHER FOREST TRANSITION drivers developed their effects in India. The Forest Conservation Act of 1980 contained provisions to reduce deforestation and land use change and set out an obligation for forest plantation to offset deforestation for infrastructure projects. But the real turning point was linked to an almost revolutionary development: Joint Forest Management.

This idea originated with a divisional forest officer, A. K. Banerjee, in the 1970s in the state of West Bengal. Banerjee, who was desperate about the festering, unproductive conflicts of the forest department with local communities, established village forest protection committees to share the benefits from forest operations with the villagers. He allowed them to collect deadwood, fruit, and nuts from the forest and offered labor employment. Local communities were also given 25 percent of the timber sales. This was a courageous initiative by an individual forest officer, as it did not conform with Indian forest laws.

By the early 1990s there were over 1,800 local forest protection committees and over 240,000 hectares of Sal (*Shorea robusta*) forest under Joint Forest Management in West Bengal alone.[13] Other Indian states followed the example of West Bengal, and in 1988 a new National Forest Policy Act introduced the principle of participatory forest management. The area under Joint Forest Management increased to 15 million hectares in 1990. It contributed in a major way to India's forest transition—a remarkable development given India's continued population growth. Between 1987 and 2003 the total Indian forest area increased by about 6 percent to cover 20.6 percent of India's geographic area. This figure, however, includes open forest areas with a crown cover of 10–40 percent.[14]

VIETNAM

SATELLITE IMAGERY OF forest cover in Vietnam throughout the 1990s shows a forest transition from 25–31 percent forest cover to 32–37 percent. This was a consequence of natural reforestation and forest plantation. However, old-growth forest was still cleared during this period, leading to lower forest density and a higher proportion of young and degraded forests.[15]

REFORESTATION: MAINLY IN MARGINAL AREAS

A RECENT COMPILATION of forest recovery studies suggested that about 23.5 million hectares (1.2 percent of the total tropical rainforest area) were experiencing a longer-term "secondary regrowth."[16] Short-term fallow areas expected to be cleared again were excluded in this compilation, and only secondary forest areas "committed" to regrowth were included. In FAO terminology (see box 2.2) this appears to be synonymous with "naturally regenerated forests," but probably includes areas of deliberate reforestation and afforestation as well. About 70 percent of these areas of regrowth were found on hill and mountain slopes less accessible for agricultural purposes, mainly in Central American and Asian forest areas that have been populated for a long time.

The TREES project estimated 50 percent less regrowth during the 1990s than this study.[17] However, these estimates are not directly comparable, as satellite imagery of deforestation hot spots may not show regrowth that occurs in the same areas, and there is no reliable method to distinguish between various forms of regrowth that may all be classified as "forest."

An extensive new study of reforestation trends across 16,050 municipalities of all forty-five Latin American countries revealed that environmental rather than demographic variables determined reforestation.[18] More than 40 percent of the woody vegetation increase during the 2001–2010 period occurred in the drier forest types of higher elevations, particularly in the dry shrub biomes of northeastern Brazil and northern Mexico. Most of the deforestation, on the other hand, was concentrated at lower elevations, in tropical rainforest areas with low population density (notably the Brazilian arc of deforestation).

Modeling forest transitions

THE U-SHAPED CURVE of forest transitions observed in temperate countries prompts the important question of what parameters govern the course of forest cover change over time. Some authors have linked forest transition to demographic developments and urbanization. Others have generalized even further, correlating forest transitions to the environmental Kuznets curve.[19] The latter postulates that with rising per capita income in a country the environmental impact increases as well, but only up to a certain point before it starts to decrease again. Many people consider the environmental Kuznets curve a flawed concept, as it does not take into account the externalization of environmental costs—when the damage done to the environment, for example through the over-exploitation of resources, is shifted to other countries.[20]

Meyfroidt et al. have shown that externalization of environmental impacts has happened in certain cases of forest transition.[21] In Vietnam since the early 1990s, a displacement of deforestation by the rapid net gain of forests was accompanied by an increase of timber imports from other countries, including from illegal sources. Figure 5.2 shows the relationship between agricultural and forest area changes and net displacement effects in four countries. The reforestation in China in combination with the 1998 logging ban had the same effect as the reforestation in Vietnam, but to a much greater extent because of the size of the Chinese market. These authors also found that the reforestation in China and El Salvador displaced the demand for crops and livestock to other countries, not least to some of the most forest-rich countries with large areas of primary forests, such as Indonesia and Brazil.

Not all forest transition countries displace their demand for forest and agricultural products abroad. A forest transition can be made without reducing agricultural land area if a country has fallow and degraded land that can be used for forest plantations or left for natural reforestation. This has been the case in Costa Rica. As globalization opens up new frontiers for intensive agriculture, older, less fertile areas are often abandoned, and forest regrowth may set in.[22] However, the effect of globalization depends on location, political history, and social context.[23] Considering how much unproductive and degraded land is available in certain countries, a long-term global forest transition cannot be ruled out, but the current dynamics of deforestation do not induce much optimism on this point.

FIG. 5.2. Changes in forest and agricultural area

Change in forest area (lower line, showing assessment points) and agricultural area (upper line) in two forest transition countries (Costa Rica and Vietnam) and two net forest loss countries (Brazil and Indonesia). Gray areas show total net displacement (quantities of traded goods in the agricultural, livestock, and forestry sectors converted into the area needed to produce these commodities in millions of hectares). Positive values show displacement, negative values show absorption, that is, net excess production to be exported. Adapted from Meyfroidt et al.[24]

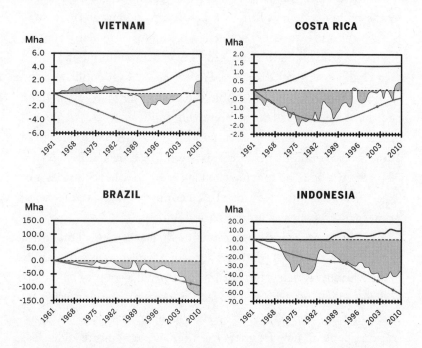

SOME HOPES FOR THE FOREST TRANSITION THEORY, BUT NO PANACEA

HOW RELEVANT ARE the European forest transition stories to tropical countries? Each country has its specific socioeconomic circumstances, which we need to take into account when we ask whether its institutional frameworks, demographics, and land use systems are conducive to forest transitions. Modeling tropical forest transitions is difficult because of the many interlinked variables. The study by Meyfroidt's group demonstrates that using generalizations like the environmental Kuznets curve or demographics to explain or predict forest transitions is not really feasible. Depending upon the local conditions, forest transitions take varying amounts of time, and there is no evidence that all or most countries will see a long-term turnaround. Forest transitions should therefore not be seen as a stage in a predictable and deterministic pattern of land use.[25]

The forest transition theory as developed by Mather in the 1990s is a relatively new discipline and does not yet have standardized terminology and assessment techniques. Because deforestation and reforestation may take place simultaneously in different parts of a country, forest change has to be carefully differentiated and classified. Forest cover may oscillate over time, so we also need longer-term studies of how land use changes in tropical countries. So many questions remain open that in their recent book, *Reforesting Landscapes*, Harini Nagendra and Jane Southworth raised the provocative question of whether forest transition theory is useful at all or whether it would be better to try to understand forest change as a cyclical phenomenon.[26]

The forest transitions seen so far in the humid tropics have raised hopes that more forest area can be regained, increasing carbon sequestration and recovering lost biodiversity, but expectations of regaining biodiversity could be totally out of place. Native species may have gone extinct with the disappearance of intact forest, and invasive species may have taken over in disturbed ecosystems.[27] Important old-growth forest may be lost even though net deforestation is decreasing—even as net reforestation is reported, as happened in Vietnam in the 1990s.

When discussing ecosystem services such as carbon sequestration, biodiversity, and hydrological services, we should distinguish between gross deforestation (loss of intact forest) and reforestation, because net deforestation measures that combine loss and gain of forest area are ambiguous in this context.[28] This does not mean that regrown forest does not have ecological value—if nothing else, it can serve as corridors between fragments of old-growth forests. But to hope for a complete ecosystem recovery with the original biodiversity borders on wishful thinking.

Forest transition research so far has paid little attention to the structure and composition of the regrown forest, the role of ecosystem services, or the potential for biodiversity conservation and rehabilitation. Nevertheless, the increasing signs of reforestation in tropical countries with the accompanying promise of ecological benefits raises the question of how governments can promote and enhance reforestation. Effective reforestation will require land use policies that build on existing socioeconomic tendencies and contribute to a plan to reduce carbon emissions from deforestation.

6

RAINFOREST BIODIVERSITY
Treasures without Price Tags

TROPICAL RAINFORESTS REPUTEDLY have the highest degree of biodiversity of all terrestrial areas, which means that global biodiversity is linked to the fate of tropical rainforests. Before talking about biodiversity we should be clear what we mean by this term. The term "biological diversity" was coined by the conservation biologist Raymond F. Dasmann in 1968. It became widely used in its abbreviated form "biodiversity" after the publication of the papers of the National Forum on BioDiversity, held in Washington in September 1986.[1] A few years later the term was defined in the Global Biodiversity Strategy of 1992[2] as follows: "Biodiversity is the totality of genes, species, and ecosystems in a region." It can be divided into three hierarchical categories: "Genetic diversity" refers to the variation of genes within a species. "Species diversity" refers to the variety of species within a region—or the species "richness." It considers also the diversity of different taxa, for example the diversity of plant and animal orders or families. "Ecosystem diversity" is the diversity of different ecosystems in a region, for instance distinct plant and animal communities.

Because of its three-dimensional nature, the term "biodiversity" is a theoretical concept. Its usefulness lies in the inclusiveness of the term, but at the same time the disadvantage is that it cannot be measured like the concentration of an atmospheric gas or the population of an animal species. This at least partly explains why "biodiversity" is a widely misunderstood and often misused term. Even scientists occasionally use it in a reductionist way to describe only one aspect of biodiversity—species diversity.

Assessing tropical rainforest biodiversity is a real enigma, considering the variety of tropical forests and the high degree of endemism—the patchy distribution of plant and animal species that occur exclusively in certain areas. We may have a good idea of the number of larger mammal and bird species in certain areas, but we know little about fish species diversity and often do not have the slightest idea of the number of invertebrate species. And this does not even start to consider the genetic variability within species.

Without far better scientific data, talking about rainforest biodiversity means using substitutes as indicators of biodiversity. Instead of trying to assess the biodiversity in tropical rainforests, some institutions and individual scientists have chosen to measure the dynamics of biodiversity loss over time by using surrogate indicators. Since there is no single comprehensive metric to monitor the state of biodiversity, the parties to the Convention on Biological Diversity, for example, agreed on seventeen different indicators to evaluate progress toward the 2010 CBD diversity targets.[3] Among those were the extent of forest area; the number, size, and representativeness of protected areas; and the status of threatened species. Two proxies used to estimate biodiversity loss through deforestation and other anthropogenic ecosystem changes will be discussed here: species extinction rates, and the dynamics of average species abundance of species populations expressed through such indices as the Living Planet Index.

Species richness: Reaching for the stars

TO ESTIMATE SPECIES extinction in tropical rainforests, we first need an idea of the baseline—species richness. Unfortunately, we do not know the number of species living in tropical rainforests even within a factor of ten. Only a fraction, perhaps 1.6–1.7 million eukaryotes—organisms consisting of cells with a nucleus, which excludes viruses and bacteria—have been scientifically recorded. But as there is no centralized database of species, some species in different collections will have been counted more than once.[4]

The tremendous species richness of rainforests has fascinated the scientific world for a long time, and has influenced the estimates of total species numbers on Earth. In the 1980s Terry L. Erwin from the Smithsonian Institution collected about 1,200 different beetle species from a single Central American tree species (*Luehea seemannii*) through fogging with pyrethrum, a natural insecticide. Many of these species were unknown to science. In a hectare of Peruvian rainforest he found 41,000 arthropod species, of which more than a quarter were beetles.[5] He found symbiotic relationships between many beetles and specific tree species, which determined the distribution patterns of these insects. Erwin's amazing discoveries prompted him to extrapolate that there could be up to thirty million insect species on Earth. It was a hypothesis, but it still triggered a fiery debate among academics. Erwin himself admitted that determining the number of species is "like reaching for the stars."[6]

In recent times the total number of species expected to live on our planet has been trimmed down on the basis of a sophisticated analysis of whole plant and animal phyla. Many taxonomists (scientists who study the classification of species) now seem satisfied with the estimate of 8.7 ± 1.3 million species.[7] Of this global total estimate, tropical arthropods have been independently estimated to number around three million, predominantly, again, beetles. In other words, two-thirds of Earth's arthropods have not yet been discovered and described—and may in fact never be discovered before they vanish forever.[8]

The legendary ecologist Edward O. Wilson estimated in 1988 that tropical rainforests, although they occupied only 7 percent of the land surface, contained more than half of the world's species, many of them as yet undescribed. The recent findings on the immense arthropod diversity of rainforests seem to confirm his "informed guess" of a quarter of a century ago. Wilson, besides being a pioneer in ecological thinking, is probably also the world's biggest expert on ants. He had himself collected forty-three species of ant from one single tree in the Tambopata Reserve in Peru—almost as many as the entire indigenous ant fauna of the United Kingdom! According to some estimates, one-third of the animal biomass in the terra firme (solid ground) rainforest areas of Amazonia is made up of ants and termites. A hectare of forest is estimated to harbor about eight million ants and one million termites.[9]

ELUSIVE TREES AND HYPERDOMINANTS

WE MIGHT THINK that trees, being less elusive than beetles living in the rainforest canopy, would be easier to identify and tally for a complete record of the tree species diversity in a region. But even that is less than certain. Estimates vary between 43,000 and 50,000 tree species.[10] The Smithsonian Tropical Research Institute recorded 644 tree species in a single hectare of the Yasuní Reserve in the Ecuadorian Amazon, about as many as in all of North America! This is presumably the highest tree species diversity ever recorded. The Yasuní National Park is also home to the only two uncontacted indigenous tribes in Ecuador, the Tagaeri and the Taromenane. In

Peninsular Malaysia a high-diversity one-hectare plot in the Lambir Hills National Park yielded 497 tree species, whereas 494 species were found in a Cameroonian rainforest plot.

In October 2013 *Science* published the results of an enormous undertaking to assess the biogeography of the Amazonian tree communities.[11] The 122 authors of this study collected stem density and species abundance from 1,170 tree inventory plots across the entire Amazon Basin and Guiana Shield. Based on these plots, they estimated a total number of 15,000–16,000 tree species in Amazonia. These included all woody plants with a minimum diameter at breast height of ten centimeters. The authors also ventured an estimation that there could be 390 billion trees standing in the Amazon Basin.

However, the most significant finding of this megastudy was the fact that only 227 tree species—the "hyperdominants"—make up about half the individual trees in the Amazon Basin. The most common species belong to three families: the Arecaceae (palms), Myristicaceae (nutmeg family), and the Lecythidaceae (Brazil nut family). The palm *Euterpe precatoria*, the most common tree in Amazonia, accounts for 1.32 percent of all trees. The hyperdominant species, as one would expect, have generally larger distribution ranges but show a preference for the most common forest types— terra firme (solid ground forest), várzea (seasonally flooded forest), white-sand forest, swamps, and igapó (blackwater-flooded forest). Only one species among the hyperdominants—*Eschweilera coriacea*, a timber tree—is dominant across the entire Amazon Basin.

Equally remarkable is that two-thirds of tree species expected to occur in Amazonia are rare and poorly known species. The rarest 5,800 species are estimated to number less than 1,000 individual trees, whereas the most common, such as the palm *Euterpe precatoria*, have estimated population sizes of more than five billion individuals! The rarest tree species in Amazonia may be endemic species that appear nowhere else with very restricted distribution ranges and could well be globally threatened.

Extinction rates: A valid indicator for biodiversity loss?

BECAUSE THE LARGE majority of species are very small animals, mainly unknown arthropods (such as insects, spiders, centipedes, and crustaceans), it is not surprising that relatively few species extinctions have been recorded. The IUCN (International Union for Conservation of Nature) estimates that over the past hundred years or so about one mammal or bird species per year has gone extinct, but who could guess the number of less conspicuous creatures that may have vanished with them? Most species extinctions happen unnoticed by mankind, silently, much as the insects and birds in Rachel Carson's *Silent Spring* vanish.

An exception is the global amphibian decline. Since the early 1980s herpetologists have observed dramatic declines in populations of amphibians around the world. In response, the Global Amphibian Assessment—an initiative of the IUCN, Conservation International, and NatureServe—was created, and its results were published in 2004. The IUCN Red List of 2013 included 489 "critically endangered" amphibian species. Almost a third—32.4 percent—of amphibians are now threatened with extinction, many of them in tropical forest areas, particularly in Latin America. As many as 159 amphibian species may already be extinct. Among these species, the case of the golden toad (*Bufo periglenes*), an endemic species in the Monteverde cloud forest of Costa Rica, became widely known. It started to become rare in 1987 and vanished completely by 1989. The Monteverde harlequin frog (*Atelopus varius*) also became extinct in the same period. Because these species were located in the pristine Monteverde Cloud Forest Reserve and these extinctions could not be related to local human activities, they raised particular concern.

Many causes for declines in amphibian populations related to human disturbance such as habitat fragmentation, disease, ultraviolet-B radiation, and airborne chemicals have been identified or suspected, but for a number of years it was not clear what could have caused the disappearance of the golden toad and the Harlequin frog in 1989. A number of research teams traced the rapid declines back to the fungus *Batrachochytrium dendrobatidis*, which causes the deadly fungal infection chytridiomycosis. It spread fast across Central America from north to south, but has also been recorded in other parts of the world.[12] Increased cloud cover caused by climate change has been suspected of causing the deadly fungus to proliferate in the Monteverde forest reserve. Now that the Global Amphibian Assessment has tracked down some causes of amphibian extinctions, there is even greater uneasiness among biologists that this could be only the tip of the iceberg of extinctions of less conspicuous species than a golden toad.

FEARS OF MASS EXTINCTIONS AND THE ROLE OF FOREST REFUGIA

THE MILLENNIUM ECOSYSTEM Assessment[13] initiated by United Nations Secretary General Kofi Annan in 2000 included contributions from over 1,000 scientists. It reproduced estimations on extinctions and concluded that current extinction rates may be up to 1,000 times higher than the rates seen in the fossil record, and could rise to ten times that with current projections of habitat loss.

In the second half of the 1980s, as Edward O. Wilson was editing *Biodiversity*,[14] deforestation rates in Amazonia were at their worst. This led Wilson to project that tropical rainforests would mostly disappear in the twenty-first century and drive hundreds of thousands of species into extinction. In the same book, Norman Myers estimated that as many as 50,000 species may have been eliminated in Brazil's Atlantic forests and Madagascar between 1950 and 1985, and that if Amazonian deforestation continued at that rate until 2000 we could lose about 15 percent of all plant and animal species.[15] These pessimistic projections were based on the theory of the species–area relationship, which assumed species extinctions will be commensurate with the loss of forest area. They also considered the patchy distribution of many rainforest species and the very large number of yet undescribed arthropods, primarily beetles.

However, another contributor to *Biodiversity*, Ariel Lugo, found that the massive deforestation in Puerto Rico, which lost up to 99 percent of the primary forest area, did not lead to the massive extinction expected by Myers et al. After 500 years of human pressure, only seven bird species became extinct, which corresponded to 11.6 percent of the indigenous bird fauna. Introduced species had even increased the number of species on the island. Lugo attributed this astonishing fact to secondary forests that had served as refugia—areas of suitable habitat—for relict forest tree species. After a few decades, some of these secondary forests started to resemble primary forests again.[16]

Similar observations were made in El Salvador, which had lost more than 90 percent of its natural forest even before the war of the 1980s and since then experienced some forest resurgence (see chapter 5). Despite massive forest cover loss, El Salvador seems to have preserved impressive levels of biodiversity, thanks to relatively small conservation areas and a variable human-influenced landscape.[17] Shade coffee plantations, where the coffee is grown under a tree canopy, in particular provided a substitute forest habitat for many bird species. Of the 508 bird species known to occur in El Salvador, 270 are habitat specialists with highly restricted ranges. At the end of the 1990s those in danger of extinction numbered 117, but only three were then believed to be extinct.[18]

These findings from Central America and the Caribbean shed some doubt on extinction rates extrapolated from the species–area relationship. They also emphasize the importance of conservation areas, undisturbed forest remnants, and mixed forest-agricultural landscapes for the preservation of certain species. Of course, this should in no way be used to downplay the much greater importance of continuous intact forest areas. But as the world's tropical rainforests become more fragmented, conservationists will also have to take the value of degraded and fragmented forest landscapes into account. Just as the value of biodiversity cannot be expressed with a single figure, so is biodiversity conservation not a simple business of "all or nothing."

NEW SPECIES GET THE HEADLINES

SMALL, OBSCURE, OR rare rainforest species disappear without anybody noticing, but the scientific and popular presses are eager to report a new species from the tropical rainforest. Every year new species of vertebrates are found in rainforests. The Vu Quang Nature Reserve in Vietnam has been particularly productive in this regard. It astonished many people that even a war-stricken country like Vietnam could conceal new large mammal species; in 1993 a bovid species previously unknown to science, the saola (*Pseudoryx nghetinhensis*) was discovered, and only a few years later another large ungulate in the same area, the giant muntjac deer (*Muntiacus truongsonensis*).[19]

The rainforests of Borneo are also a haven for elusive vertebrate species. A few years ago a new species of slow loris (*Nycticebus kayan*) was detected—with the added peculiarity that it was found to have a toxic bite! In the largest remaining rainforest area of the world, Amazonia, at least 441 new species were discovered between 2010 and 2013 according to Cláudio Maretti, leader of the Living Amazon Initiative at WWF. These discoveries were the result of collaborative research in remote areas of Amazonia by scientists from around the world. The new species consisted of 258 new plant species, 84 fishes, 58 amphibians, 22 reptiles, 18 birds, and 1 mammal. It was the mammal that attracted the most curiosity, as it purrs like a cat when it is at ease; the purring monkey (*Callicebus caquetensis*) is now also called the Caquetá titi monkey after the Caquetá region in the Colombian Amazon. It belongs to a group of about twenty different species of titi monkeys that occur only in the Amazon rainforests.

Whether it's the saola or a purring monkey, every time the tropical rainforests of the world unveil a previously unknown mammal or bird, it becomes a widely reported news event, as a quick Internet search for the saola demonstrates. The excitement over newly detected species promotes an utterly distorted picture that there could be more species appearing than disappearing. The *Economist*, for example, when referring to the extinction rates mentioned in the Millennium Ecosystem Assessment, wrote that "nobody now thinks that anything remotely on that scale has happened."[20] Although such reporting may be nothing more than an attempt to discredit science-based models, it reflects how little we know about the extinctions of smaller species, to say nothing of thus far undescribed invertebrates.

WITHOUT A BASELINE, EXTINCTION RATES REMAIN HYPOTHETICAL

IF WE ASSUME the global extinction rate specified in the Millennium Ecosystem Assessment, 1,000 to 10,000 times faster than the background rate we see in the fossil record, then between 0.01 and 0.1 percent of all species would go extinct each year. Based on the estimate mentioned earlier of 8.7 million species on our planet, this would mean between 870 and 8,700 species would be lost each year.

People have translated such guesses into species lost per day, which then prompts the silly question: "So which ones did we lose yesterday?" As ridiculous as such a question may sound, it is symptomatic of the light-years that separate the ease of, for instance, buying and selling shares on the stock market in a fraction of a second on the one hand and the disheartening scientific and biophysical obstacles to understanding life on Earth on the other.

Extinction rates, at least for tropical rainforests, remain hypothetical. Not only do we have no idea how many organisms exist in a forest area, but we also have no reliable data on distribution patterns for most species. In other words, there is no baseline from which to calculate extinction rates.

Even with accurate baseline information, who would be able to survey the hundreds of thousands of arthropods to see whether a spider or beetle species was missing in the crown canopy of a certain area, and since when? Analyzing historical extinction rates of well-known species groups in a well-defined geographical area and projecting them into the future is not impossible. The scientifically demanding forest fragmentation studies discussed in this chapter do just that. But as a surrogate of biodiversity loss, extinction rates that are not based on earlier scientific assessments of species numbers and their populations remain hypothetical and are far too uncertain to be of any real use.

Species population trends point to massive biodiversity loss

INSTEAD OF USING species extinction rates to estimate biodiversity loss or gain, some institutions have developed indices-based variations in species' populations. The mean species abundance index (MSA), for example, is assessed as the average abundance of species, compared to their abundance in a pristine state of the ecosystems they inhabit. The MSA is computed on the basis of the GLOBIO model, developed by the Netherlands Environmental Assessment Agency, UNEP/GRID-Arendal, and UNEP-WCMC (World Conservation Monitoring Centre).[21] This model uses map-based spatial information on environmental drivers, such as changes in land use, infrastructure, or climate, to simulate impacts on terrestrial biodiversity over time. Thus the MSA index is a simulation that uses no detailed species data. According to the MSA, about 27 percent of the original global biodiversity had been lost by 2000, with a further 11 percent projected to be lost by 2050 in a business-as-usual scenario.

THE LIVING PLANET INDEX

UNLIKE THE MSA, the Living Planet Index (LPI) uses data compiled since 1970 about actual populations to monitor changes in population size for vertebrate species from different biomes and regions. The LPI was developed by WWF in collaboration with UNEP-WCMC and has been published every other year since 1998 as part of the *Living Planet Report*.[22] It is computed in collaboration with the Zoological Society of London. The 2014 edition of the LPI was based on over 10,000 vertebrate populations belonging to 3,038 mammal, bird, reptile, amphibian, and fish species. The global Living Planet Index showed a decline of 52 percent between 1970 and 2010. This is a steeper decline than reported in previous years when the data from North America and Europe dominated too much. The decline is also stronger than the mean species abundance index suggests. However, the two indices cannot be compared directly, as the MSA is a simulation based on a hypothetical pristine state until 2000, whereas the LPI monitors population data between 1970 and 2010. Vertebrate populations included in the LPI are classified according to whether they occur in temperate or tropical regions, and whether they live in terrestrial, freshwater, or marine ecosystems.

Terrestrial and freshwater species populations were also classified into biogeographic realms. For the purpose of this book, the three tropical realms are of particular relevance: the Afrotropical, the Neotropical, and the Indo-Pacific. The latter two realms show dramatic declines since 1970, whereas the LPI for the Afrotropical realm documents some recent increases in population levels (mainly birds and fish species). This leads to greater variability of the index (see figure 6.1).

FIG. 6.1. The Living Planet Index

Accumulated trends of terrestrial and freshwater vertebrate populations of biogeographic realms between 1970 and 2010. The tropical index shows a steeper decline than the temperate one. The Neotropical and the Indo-Pacific living planet indices show particularly drastic declines. Shading surrounding the trend represents the 95 percent confidence limits. The wider the shading, the more variable the underlying data. From *Living Planet Report* (2014).[23]

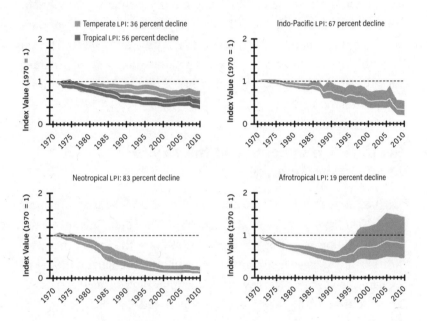

As the index is set at 1 in 1970, declines in species populations that occurred before 1970 are not accounted for in these statistics, which means that the important biodiversity losses of temperate zones that took place in historic times, before 1970, are not displayed by the LPI. The recent loss in tropical regions could be considered to be a similar price of development as the one paid earlier in temperate regions.

Not all the population data used to calculate the tropical LPIs are from tropical rainforest ecosystems. Nonetheless, the continued, alarming declines, particularly of the Neotropical and Indo-Pacific LPI, show the result of deforestation and forest fragmentation. The decline of these two indices gives rise to serious concerns that with continued deforestation and forest fragmentation, many vertebrate populations could soon become nonviable or go extinct altogether. Although the decline of the regional LPIs is clearly linked to deforestation and forest fragmentation, the relationship is certainly not a linear one. Species react differently to the various forms of habitat change, and natural fluctuations may play a role as well. The LPI is a cumulative measure, but with the advantage that it provides an indication of biodiversity trends based on verifiable data, rather than simulations.

Rainforest fragments and the theory of island biogeography

IN 1967 ROBERT H. MacArthur and Edward O. Wilson first published *The Theory of Island Biogeography*.[24] This book advanced the understanding of ecology and evolution more than any other publication in the past half century. It used a mathematical model to explain the species diversity in a geographically confined area, such as on an island. The theory also demonstrated that the equilibrium between immigration and extinction of species is a function of the size and distance of an island from the mainland, or the main population of the species. MacArthur and Wilson found that among islands of different sizes, the number of species that each island can support increases approximately with the fourth root of the area of the island. Thus in a typical case, an island ten times the size of another can harbor close to twice as many species. Forty years after the publication of the seminal MacArthur-Wilson book, a conference at Harvard University reviewed and extended the current knowledge of island biogeography.[25]

Island biogeography theory is particularly relevant to the future of the tropical rainforests for two reasons: because of the restricted distribution ranges of many species within the forest, and because of the increasing fragmentation into forest "islands" in many countries. Robert MacArthur in his last book, *Geographical Ecology*,[26] published in 1972 (he died of cancer the same year at the age of forty-two), explained the theoretical basis for the patterns of species distribution we observe.

MacArthur reflected on the patchy distribution of rainforest species, especially bird species, observed by the remarkable physiologist and evolutionary biologist Jared Diamond. In his student years Diamond specialized in bird species distribution in New Guinea, many years before he published the now famous books *Guns, Germs, and Steel* and *Collapse* among others. MacArthur presented four explanations for the patchy distribution of rainforest species:

1) historical gaps in suitable habitat are still mirrored in the species distribution today, even though the habitat is now continuous;

2) competition from a similar species keeps the gap area clear;

3) factors that are not easily recognizable make the habitat in the gap area unsuitable; or

4) the patches are remnant populations of a species going extinct.

These explanations are not entirely independent from each other, but MacArthur implied that since competing species are packed closer together in the tropics, these subtle differences in local conditions could cause the patchiness.

FOREST FRAGMENTATION: A GREAT RISK FOR BIODIVERSITY CONSERVATION

IN VIEW OF the increasing fragmentation of tropical rainforests, especially in Southeast Asia and in the southern states of the Brazilian Amazon, to what extent can island biogeography theory predict the biodiversity impact of fragmentation? One of the contributors to the forty-year review of *The Theory of Island Biogeography*, William F. Laurance from the Smithsonian Tropical Research Institute, tested the theory in fragmented tropical forest areas in the Amazon. He found that while the theory is relevant in principle, it fails to take into account factors besides size and isolation. Other important factors that affect the vulnerability of species and their chances for survival are edge effects and such human-induced perturbations as selective logging, overhunting, and climate change.[27]

Laurance's data came from a long-term research program on forest fragmentation initiated by Thomas E. Lovejoy in 1979 in a terra firme forest area north of Manaus in central Amazonia. In 2011 a summary of the remarkable thirty-two-year investigation was published.[28] It provided extremely important insights into the dynamics of fragmentation and its impact on biodiversity. The Biological Dynamics of Forest Fragments Project is the world's largest and most relevant research program on tropical forest fragmentation. As of 2010 it had produced 562 publications and 143 graduate theses.

FRAGMENT "ISLANDS": THE AREA MATTERS

ONE OF THE primary goals of the forest fragments project was to understand the relationship between the size of the fragments and the species richness—a classic question of island biogeography. Forest fragments had a size of between one and a hundred hectares. Many animal and plant groups were monitored in these fragments and their status was compared with their situation before fragmentation.

It was found that over time, species richness declines with fragment size. Declines were especially steep for primates, herbivorous mammals, and insectivorous understory birds (see box 6.1), but also for tree seedlings and palms. Smaller fragments tended to lose species more quickly, as would be expected from island biogeography theory. The investigators estimated that even fragments of 10,000 hectares would lose some of their bird species over a century.

Tree felling by a traditional West African shifting cultivator. (Photo by C. Martin.)

As tropical rainforests typically show a patchy distribution of species, many of which are rare, the "sample effect" also becomes relevant to the viability of isolated forest areas.[29] The sample effect suggests that some species may be absent in fragments not because they became locally extinct, but because they were already absent when the fragmentation occurred. This effect is particularly marked when it concerns plant and animal species with typically patchy distribution patterns.

BOX 6.1

The theory of island biogeography mirrored in forest fragment dynamics

The theory of island biogeography described by Robert MacArthur and Edward O. Wilson in 1967 was undoubtedly an important force behind the long-term fragmentation studies undertaken by the Biological Dynamics of Forest Fragments Project. However, in recent years research started focusing increasingly on other aspects besides the size of fragments and their isolation. Parameters such as fringe effects, overhunting, climate change, and the influence of the matrix—the vegetation in surrounding areas—were also found to be important for predicting extinction.

A twenty-five-year project that monitored understory bird communities nevertheless revealed a dominant influence of the fragment size on extinction rates (see figure on page 153).[30] The most extreme case of a fragmentation-sensitive species is the wing-banded antbird (*Myrmornis torquata*), which was present in all eleven fragments before those were isolated but had disappeared from all of them twenty-five years later. Surprisingly, about half the species that went extinct in the earlier part of the project recolonized some of the fragments in the latter part, and this happened in the fragments of all sizes. Generally the species diversity declined in the first ten years after isolation but leveled off after that. Although species continued to go extinct, they were replaced by recolonization in the latter part of the twenty-five-year observation period.

Bird species extinction in forest fragments

Percentage of understory bird species that went extinct in eleven Amazonian forest fragments of one hundred, ten, and one hectare. The time span is from the early 1980s, before isolation, to 2007 (about twenty-five years). Adapted from P. C. Stouffer et al. (2011).[31]

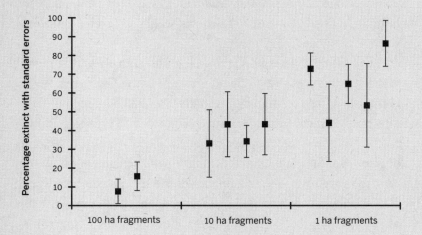

EDGE EFFECTS AND SURROUNDING VEGETATION

FRAGMENT SIZE IS not the only factor that influences species richness over time: the Biological Dynamics of Forest Fragments Project documented an important influence of forest fragment edges with regard to hydrology. Desiccation in surrounding areas draws moisture from forested areas, and increased wind shear and turbulence increases tree mortality. Many insect groups and smaller understory birds are sensitive to these edge effects, which can even occur along some of the wider forest roads.

The type of vegetation surrounding the fragments (matrix vegetation) influences the species abundance in the fragments. Forest regrowth in adjacent areas generally diminishes negative edge effects, and fragment edges may overgrow with creepers and lianas, "sealing" the fragment against outside climatic effects to some degree. The plant and animal communities in fragments, however, tend to move toward a resemblance of the species composition in surrounding areas. Whereas forest-dependent species may use forest regrowth to recolonize fragments, clearings of 100–200 meters wide, on the other hand, become effective barriers for many species. Cattle pastures have a particularly negative effect on the movements of forest animals. The effect of the matrix vegetation on the forest fragments is similar for all the fragments in a given landscape, but the dynamic of species extinctions and recolonizations may differ greatly between landscapes with different matrix vegetations.

IMPORTANT LESSONS FOR BIODIVERSITY CONSERVATION

THE AUTHORS OF the huge Biological Dynamics of Forest Fragments Project found a "bewildering variety of ecological distortions" that affect fragmented habitats. Understanding these ecological distortions is important for designing protected areas and planning land use for biodiversity conservation. In the 1970s and 1980s a scientific debate raged over whether biodiversity was conserved better by a single large reserve or several small reserves with the same total area. At least until the sample effect—the random inclusion or exclusion of species with a patchy distribution in a sample (a forest fragment)—became better understood, there was no consensus on the question of "single large or several small."

The forest fragments project brought further proof that habitat fragmentation is a major factor leading to loss of global biodiversity. Besides biodiversity loss, fragmentation affects biomass and carbon storage, as many trees tend to die near forest edges. The thirty-two-year investigation concluded that nature reserves in Amazonia should ideally be in the range of thousands to tens of thousands of square kilometers to provide sufficient security for the survival of the many rare species and species with a patchy distribution. Reserves should also cover a variety of different river basins as well as climatic zones and soil types.[32]

The insight into the dynamics of fragmentation provided by the Biological Dynamics of Forest Fragments Project and the findings of the research in Central America on the role of forest regrowth[33] started to move the discourse on rainforest biodiversity away from a mere focus on total forest cover. For the first time some light was shed on the complex parameters of local species extinctions in forest landscapes, as bewildering as they may still appear. Long-term studies, particularly of bird diversities in South American forests, have previously shown drastic declines of forest-dependent species in fragmented forest areas. However, these studies were mostly based on checklists—the counting of occasionally observed species over varying time periods—which are known to be unreliable indicators.

The results from the forest fragments project show that biodiversity is influenced by factors besides size and isolation. An extremely important aspect is increasing hunting pressure in fragmented forest areas, particularly in Africa where bushmeat consumption is a deeply rooted tradition and an important part of local diets. In West and Central African countries bushmeat is not just eaten locally, but sold in urban centers and exported to other markets. The expanding areas of "empty forests" where the larger mammal species have been overhunted and poached even in protected areas show that overhunting is a more serious threat to biodiversity in many regions than the reduction of absolute forest area (see box 6.2).

The analysis of the patterns of local extinctions in disturbed and fragmented forests has shown that biodiversity conservation strategies should also include areas outside large primary forest areas. The research cited above emphasized that small forest remnants should be protected and that secondary vegetation and forest regrowth play an important role as bridges connecting forest remnants.

BOX 6.2   Empty forests: The price of fragmentation

A forest is a vastly diverse community of living organisms that depend on each other in an intricate system of thousands of relationships. Many plants depend on animal species as pollinators or seed dispersers and share a coevolutionary history. In the 1970s, when I lived in the rainforest areas of the Western Region of Ghana and was in charge of the Bia National Park and a number of other reserves, my park rangers and I collected data on the phenology (the periodic cycles of flowering and leaf-flushing) of forest trees, forest elephant movements, and primate populations to help us understand some of these relationships.[34]

There were eight higher primates species in Bia National Park at that time. In a study area of less than seven square kilometers in the center of the park, we recorded and mapped the home ranges of the three top-canopy-dwelling species: two groups of red colobus (*Colobus badius*), the larger of which counted fifty-eight individuals in 1978, six groups of black-and-white colobus (*Colobus polykomos*) of between ten and twenty animals, and three groups of diana monkeys (*Cercopithecus diana*) with up to nineteen individuals. To avoid food competition, the three species shared the forest habitat in a complex pattern of vertical and horizontal spatial niche separation and had different food habits. Their diet was a varying mix of leaves, fruit, and seeds belonging to a total of 50 tree species in the case of the red colobus and over 150 tree species in the case of the diana monkey.

Watching a large group of red colobus moving in gigantic leaps from one tree crown to the next in single file, constantly communicating with grunts and shrieks, was an unforgettable experience. It also showed why this primate species was the most vulnerable to hunting. Because they were easier to spot than species inhabiting the lower, denser layers of the forest, they could be shot more easily, one after the other when moving in a single line through the forest canopy. The red colobus is now considered to be extinct in Ghana and most probably as a subspecies.[35] A recent survey carried out in the Dadieso Forest Reserve, a pristine forest area south of Bia National Park, tried to establish whether the diana monkey still survives in that area. The researchers did not see or hear a single higher primate on more than 130 kilometers of transects.[36] This result confirmed their worst fears for the fate of the diana monkey and also for other primate species.

After the 1970s the protected areas in the tropical rainforest areas of Ghana became increasingly isolated when timber concessions were allocated outside forest reserves. The small-scale logging enterprises opened up the forest to slash-and-burn farming. Throughout the 1980s and 1990s the previously contiguous forest areas fragmented, and hunting pressure increased dramatically. What seemed inconceivable some forty years ago is now a reality: the remaining rainforests of Ghana are largely "empty forests" depleted of their previous wealth of larger vertebrate species. It is interesting to note, though, that the remaining forest areas correspond almost exactly to the system of state forest reserves and national parks (see plate 10). Thus, Ghanaian authorities have done a remarkable job of preserving the legally protected forest areas from deforestation since colonial times, when these forests were classified.

Although timber exploitation continued in harvesting cycles in the forest reserves, they hardly suffered any slash-and-burn farming from migrant farmers, as happened in practically all forest reserves (forêts classées) of Côte d'Ivoire. But with overhunting and the bushmeat trade, the biodiversity value of reserved forest and parks is now greatly diminished. Bushmeat, which has traditionally played an important role in the diet of rural African communities, has become a traded good both for its food value and because it is considered a delicacy by urban populations, even on other continents.

With the absence of many seed-dispersing animals, the plant diversity is equally in jeopardy, even if the forest appears to be intact. Many tree species, including timber trees, depend on such seed dispersers

as elephants, monkeys, and ungulates. In a recent study in southeastern Nigeria, forest areas with high hunting pressure were found to have
significantly fewer tree seedlings of primate-dispersed tree species.[37]
The forest fragmentation in West African countries is a stark example of how fragmentation decreases biodiversity. It should serve as a
timely warning signal for the Central African forest countries that still
have large unbroken forest areas.

Red colobus in Bia National Park in 1977

Today this primate species is extinct in Ghana—a victim of overhunting.
(Photo by C. Martin.)

We should not let a forest full of trees fool us into believing all is well.
KENT H. REDFORD[38]

The economics of ecosystems and biodiversity

WHEN THE UNITED Nations Millennium Ecosystem Assessment was released in 2005, among its other findings, it promoted the concept of "ecosystem services" to policy makers and the business community. This concept fueled an emotional argument about monetizing nature. An article under the title "Selling Out on Nature" expressed a commonly held position as old as the conservation movement itself: that nature had to be conserved for nature's sake, and conservation should not be motivated by an anthropocentric and utilitarian view of the natural world.[39] But in the meantime "The Stern Review on Economics of Climate Change"[40] had catapulted the economic aspects of a human-induced global threat into the minds of the world's decision makers. The time had come to think of the next global threat: biodiversity loss.

In March 2007, environment ministers from the G8+5 countries meeting in Potsdam, Germany, started to analyze the global economic benefit of biological diversity, the cost of loss of biodiversity and ecosystem degradation, and the consequences of failing to take effective conservation measures. The comprehensive study "The Economics of Ecosystems and Biodiversity" (TEEB) was led by the economist and former banker Pavan Sukhdev, who posted the following message on his personal website in the first days of 2013:

> They were right, the world has ended. The world as we knew it came to an end more or less when the Mayan calendar ran out, on winter solstice, 2012. The old world that died in 2012 had a stable climate, cheap commodities, governments managing change, high growth in output and consumption, and corporations driving economies and serving societies all deeply infused with a blind belief in free markets.

The new world that begins now, in 2013, will be defined by frightening climate instability, commodity and food prices ratcheting skywards, impotent governments reduced to spectators in their own countries, continuing recession, and corporations either being hounded as common criminals or laurelled as champions of virtuous change. An Internet-enabled anarchy of the wronged will pull entire countries along unplanned roller-coaster rides through a Disneyland world of commotion and crises...

Our world is rapidly approaching planetary boundaries—across climate, biodiversity, nitrogen, [phosphorus,] ocean acidification, freshwater scarcity, inter alia. Economies worldwide are still headed in the wrong direction—towards resource exhaustion, social disparities, and persistent poverty. What we need is change at the speed of light, driven by bold leaders. What we have is change at the speed of change, hesitantly nudged by cautious governments. Too little, too late, such prevarication will spill into natural disasters. We saw some as 2012 ended—we shall see more as the earth's ecosystems do what all systems in equilibrium try their best to do: stay in equilibrium, until they simply cannot. Until what we call "resilience" is replaced by "thresholds" being breached, planetary boundaries being crossed. Breached thresholds will lead ecosystems into new states of equilibrium, which may not be any good for human life, society, or economy.[41]

The TEEB report aimed to promote a better understanding of the economic value of ecosystem services. By offering economic tools that take this value into account, it hopes to prevent further loss of biodiversity. One of the major results of this work was that it demonstrated the need to do cost–benefit analysis before making policy decisions to make sure policies are as effective as possible. The TEEB project recognized four groups of "ecosystem services":

1) *provisioning services* are the outputs from ecosystems, including food, water, raw materials, and medicinal plants;
2) *regulating services* include the control of climate and air quality, carbon sequestration and storage, erosion control and soil fertility, pollination, and biological disease control;
3) *habitat and supporting services* provide habitat for species and maintain genetic diversity; and
4) *cultural services* provide recreational, aesthetic, and spiritual benefits.

THE COST OF LOST ECOSYSTEM SERVICES

PRELIMINARY FINDINGS OF the TEEB report were presented as an interim report at the Ninth Conference of the Parties (COP 9) to the Convention on Biological Diversity in Bonn in May 2008. The interim report described, among other things, a project by a consortium led by Alterra of Wageningen University.[42] It attempted to assess the cost of not making policy changes to prevent biodiversity losses and environmental damage. The assessment was based on the Organisation for Economic Co-operation and Development's Environmental Outlook to 2030 on the future economic and demographic development. Between 2000 and 2010, biodiversity was lost that could have produced EUR 50 billion worth of ecosystem services per year. If there is no action to save biodiversity, the cost of lost ecosystem services would increase to EUR 14 trillion per year by 2050. The majority of this loss resulted from the shrinking of the tropical forest biomes.

The role of tropical forests in providing ecosystem services was again demonstrated in a recent study of the TEEB Natural Capital Coalition. It estimated the financial risk from unpriced natural capital inputs, such as the availability of clean air and water, to primary production and processing.[43] The study ranked the top one hundred environmental impacts globally. The cost of land conversion in South America for agriculture, particularly cattle ranching, was estimated at USD 312.1 billion (in 2009 prices), second only to the greenhouse gas impact of coal power generation in Eastern Asia. Even more striking was that the annual natural capital cost (not accounted for in conventional profitability calculations) of land conversion was 18.7 times the revenue generated from the same area of land. This ratio is hardly surprising if one considers the very low cattle stock density—often no more than a single cow for a hectare of cleared tropical forest.

IS ASSIGNING DOLLAR VALUES TO ECOSYSTEM SERVICES SACRILEGE?

THE TEEB PROJECT produced a background report on the ecological and economic foundations, several reports for specific groups of decision makers (national and international policy makers, local and regional policy and management, and business and enterprise), and a synthesis report. The final results of these reports were presented at the Convention on Biological Diversity (COP 10) in 2010.

At this meeting, the focus on economics in these reports again triggered the fears of the more traditional wing of conservationists and the deep ecology movement. They felt that nature was being monetized, and that its intrinsic value and the right of every species to survive would be ignored in favor of economic logic applied, ironically, by the very people who carried the heaviest responsibility for the ruthless exploitation of nature. Even reputable economists called the environmental cost–benefit analysis a "monistic pseudo-economic discourse" designed to appeal to business and governments.[44]

A purely economic view of the value of ecosystems and biodiversity is certainly reductionist and anthropocentric. And the TEEB project does emphasize the monetary effects of ecosystem loss, even though it claims to give attention to the underlying ecological and sociocultural values. But this discussion about the ethics of giving nature a monetary value is beside the point. For decades we have witnessed colossal worldwide biodiversity loss and ecosystem degradation, and it is caused by the dominant economic model. It is driven by resource extraction, production, and consumption that we measure with the deficient yardstick of GDP.

These are horrendous market failures, and we can only tackle them by going to the root causes of what is wrong with classical economic models. There is nothing immoral about seeing biodiversity loss through the lens of economic and business risk if it leads to real cost accounting and policy decisions that preserve ecosystems and biodiversity. For far too long, economists have worked in a world of their own, untarnished by an understanding of the limits of our planet. To them, the arguments of scientists and environmentalists have long sounded like an alien language without relevance to the economic and financial world. We shouldn't complain when an age of economic enlightenment starts and economists talk to economists about these issues in their own tribal language.

CAN GROWTH SAVE BIODIVERSITY?

IT IS AN equally encouraging sign if the Economist devotes a special report to the importance of biodiversity.[45] The magazine has not always recognized environmental issues as worthy of their editorials and still refers to conservationists in rather derogatory terms as "the greens." More disturbing is that the standard economic growth model that permeates almost all the articles in this special issue is portrayed as the solution to biodiversity loss and the way to prevent the next extinction: "More growth, not less, is the best hope for averting a sixth great extinction" insists the Economist. Even the environmental Kuznets curve, which has long since been shown to be flawed, is reanimated for the purpose of liberal economics (see chapter 5). And so the planting of trees in China is uncritically attributed to the country's growth. It conveniently omits mentioning the logging ban introduced after the Yangtze floods of 1998 and the massive timber imports, not least from illegal sources in Russia and Indonesia that it triggered.[46]

The problem with the liberal growth paradigm is that it does not account for the environmental damage and loss of biodiversity caused in other parts of the world. Industrialized countries and some of the emerging economies rely heavily on natural resources such as beef, soy, and palm oil, which are causing deforestation hot spots in Brazil and Indonesia.

In its special report, the *Economist* uses biodiversity conservation to support a growth doctrine that is just as flawed as the opposite claim: that a country has to be poor to keep its biodiversity. The environmental movement does not see economic growth and technology as enemies of biodiversity, as the report claims. The TEEB report, fortunately, stood clear of any economic ideology when proposing measures to safeguard biodiversity and ensure ecosystem services.

MORE THAN PRICE TAGS

THE REAL PROBLEM with the Stern Review as much as the TEEB report is that monetizing the value of natural capital and ecosystem services does not in itself lead to policy changes, and neither does the calculation of the natural capital costs of production and consumption. Policy changes happen through far-sighted governments, political will, and a functioning multilateral system to set the regulatory framework for sustainability. As the critical economist Clive L. Spash put it: "Changing the international banking and financial institutions to redirect development away from environmental destruction would seem to require a little more than making wild claims for the monetary value of bees."[47]

As biodiversity loss happens at a local scale in a myriad of places, the drivers of degradation also vary tremendously. Without understanding these drivers, we cannot translate an ecosystem services valuation or a policy change into sustainable use. The complex pattern of tropical deforestation is particularly challenging. As we have seen from the TEEB Natural Capital Coalition's study, "Natural Capital at Risk," the ratio between the high natural capital cost of deforestation and the very low revenue generated from land use change is enormous. But how can this valuation become relevant to a poor subsistence farmer in the middle of the Congo Basin, far off the radar of any government control, when he has no alternative to making a living from an intact piece of forest? This question, of course, is also at the heart of the difficulties facing REDD. The economics of ecosystems and biodiversity may be a useful tool for policy makers and to some of the more enlightened governments. To what extent it can be made relevant to mitigating tropical deforestation and to the livelihoods of the hundreds of millions of rural people who depend directly on these forests has yet to be seen.

DAVID KAIMOWITZ

Indigenous Peoples and Deforestation in Latin America

Indigenous peoples have traditionally managed a large portion of the forests in Latin America, and these forests have generally suffered lower rates of deforestation than other forests. It is important to understand why this is, whether it is likely to continue, and what might be done to ensure that it does continue.

In examining these questions, I have concluded that a combination of cultural, economic, and demographic factors together with government policies contribute to lower deforestation rates in indigenous areas, but there is a growing threat of encroachment by powerful outside groups. To reduce deforestation and forest degradation with their associated greenhouse gas emissions, government policies should strengthen indigenous land and forest rights and support indigenous authorities, and international groups should provide funding for indigenous organizations and land management.

Indigenous ownership and management of the forests of Latin America

There is no one simple definition of who the indigenous peoples of Latin America are. I use "indigenous" here to refer to Amerindian populations that have maintained important elements of their pre-Hispanic cultures. This includes hundreds of different ethnic groups, each with its own culture, traditions, institutions, and production systems. The term also covers a range of situations, from groups that have retained their languages and many of their most important cultural traits to others that have gradually lost them.

I will focus here on the lowland indigenous peoples in tropical rainforest regions and the indigenous peoples who live in the coniferous forest regions of Mexico. Those are the indigenous groups that manage the largest areas of forest. Indigenous peoples and other local communities formally own or have legal rights to manage more than 270 million hectares of forest in Latin America—almost 40 percent of the total forest area.[1] Indigenous peoples have also claimed significant areas of forest that they do not yet have formal rights to.

If one looks at the Amazon Basin in particular, over a quarter of the forest is in indigenous territories that have formal government titles. Once again, this does not include a substantial area of forest that indigenous peoples have claimed but have not yet gotten title for.[2] In Mexico, local communities own roughly 70 percent of the country's forests.[3] Indigenous peoples also own or manage a majority of the forests in Nicaragua and Panama,[4] and significant portions of the forests of Belize, Costa Rica, Guatemala, and Honduras.

Indigenous peoples have managed large areas of forest in Latin America for thousands of years. However, formal government recognition of indigenous ownership of most of these forests is relatively recent. Most indigenous territories received their titles in the last forty years, and many got them only very recently.[5] The simple fact that indigenous territories make up such a large part of Latin America's forests makes them important in any analysis of deforestation in the region.

Deforestation and forest degradation in indigenous territories

In general, deforestation rates appear to have been significantly lower in indigenous territories than in other forests outside protected areas. Recent studies from Bolivia, Brazil, Colombia, Honduras, Mexico, Nicaragua, Panama, and Peru all support this conclusion.[6] I was only able to find one study—from the Ecuadoran Amazon—that found that indigenous communities outside protected areas had similar deforestation rates to neighboring nonindigenous communities.[7]

A number of studies have looked at whether indigenous territories have lower deforestation rates than nonindigenous protected areas.

Here the evidence is less conclusive, but it is clear that in many cases indigenous forest management has proven at least as effective at conserving forests as traditional protected areas. The only study that examined this issue in all of Latin America found that indigenous territories were twice as effective as other forms of protection at reducing forest fires.[8] However, studies in the Peruvian Amazon and the Guiana Shield region of Colombia found the opposite: while both indigenous territories and strictly protected areas had much lower levels of deforestation than other regions, forest loss was lower in the protected areas.[9] In Brazil, it seems that in regions with low deforestation pressure, strictly protected areas prevent forest loss more effectively than indigenous territories, but in regions with high deforestation pressure the opposite is true.[10] In the case of the previously mentioned study of the Ecuadoran Amazon, the researchers discovered that indigenous territories that also had protected area status had lower deforestation rates than other protected areas.[11]

Both indigenous and nonindigenous groups can cause deforestation and forest degradation in nonindigenous territories. I am not aware of any rigorous studies that look at the relative weight of the forest loss caused by these different groups. However, anecdotal evidence suggests that nonindigenous groups and logging companies have been responsible for a large percentage of the forest destruction and degradation in indigenous territories, and that this threat seems to be growing as the agricultural frontier spreads farther into the forest.

Reasons for lower deforestation in indigenous territories

Why are indigenous territories less deforested? There are a number of possible explanations, with different implications for future trends and policy options.

LOWLAND INDIGENOUS TERRITORIES ARE INACCESSIBLE AND HAVE POOR SOILS FOR AGRICULTURE

Historically, European influence in Latin America expanded most rapidly in environments that were favorable to intensive agriculture or that had high concentrations of valuable minerals. The poor soils and high prevalence of malaria in most lowland tropical areas made these regions much less attractive to the Europeans, and that helped protect both the forests and their traditional inhabitants. This situation also reduced the incentives to build roads and transportation infrastructure in these regions, which further helped discourage the invasion of indigenous territories and the destruction of their forests.

A large body of research shows that locations that have more favorable conditions for agriculture and better access to markets tend to have much higher deforestation rates,[12] so it should come as no surprise that indigenous territories that lack these conditions have less deforestation. In recent decades, this situation has begun to change. New technologies have diminished the prevalence of malaria and made it easier to produce cattle and annual crops profitably in the lowland tropics.[13] New roads and bridges have opened up large new areas of forests to land speculation and destruction. Clearly this has increased the threats to the indigenous territories and their forests. Even so, the inaccessibility and poor soils of the indigenous territories continue to limit the advance of the agricultural frontier. That being said, inaccessibility and poor environmental conditions for agriculture are definitely *not* the only reason that indigenous territories have lower levels of deforestation. A number of the previously mentioned studies compared forest loss in indigenous territories to other locations with similar environmental conditions and access to markets and found that the former still had substantially less deforestation.[14]

INDIGENOUS PEOPLES HAVE MORE ENVIRONMENTALLY FRIENDLY CULTURES AND PRODUCTION SYSTEMS

If we ask indigenous people why their forests are better preserved, many will answer that it is because their cultures have greater respect for nature. This hypothesis is difficult to test, particularly on a large scale. Given the great diversity among indigenous peoples' cultures, histories, and relations with other ethnic groups, one might expect they would also have rather distinct land use patterns. We might also expect that as indigenous groups become more integrated into markets and more influenced by Western cultures, they might tend to behave more like their nonindigenous neighbors.

To date, most research on these issues has focused on specific regions or ethnic groups. The findings suggest that different ethnic groups often do have rather distinct production systems and land uses, even when they live in similar ecosystems and have similar access to markets and information.[15] To the extent that we can generalize, probably the single characteristic that most distinguishes indigenous and nonindigenous land use patterns is that cattle have historically played a much greater role in nonindigenous production systems than in indigenous territories. That has profound implications for deforestation, since the majority of forest clearing in Latin America is to establish pastures.

One might also expect that groups that have lived for generations in forested regions would be more likely to develop production systems where hunting, fishing, and collecting forest products play an important role than groups that evolved in environments with limited forest cover. For traditional forest dwellers, forests were not simply wastelands waiting to be tamed, but their very source of sustenance. More recently, a growing number of indigenous communities in Mexico and Honduras have managed large areas of forests for commercial production, which has also given them strong incentives to maintain those forests. To a lesser extent, this has also happened with lowland moist

tropical forests in Maya communities in the Yucatán Peninsula and
Emberá communities in Panama. So while it is certainly not a matter of
simply saying that indigenous peoples respect nature more than other
ethnic groups, there are reasons to believe that the production systems
of many lowland indigenous peoples tend to conserve more forest than
those of other ethnic groups.

INDIGENOUS PEOPLES LACK SUFFICIENT LABOR, CAPITAL, OR POLITICAL POWER TO DEFOREST LARGE AREAS

Most forested indigenous territories have relatively low population
densities, particularly in the Amazon Basin. Even though many of these
indigenous peoples have high birthrates, their populations will remain
relatively small for decades. Most of them live in extreme poverty and
lack sufficient capital to cultivate large areas or raise large herds of
livestock. Traditionally they have had very limited access to agricultural
credit or government subsidies for production or infrastructure. This
has greatly limited both their ability to clear large areas of forests and
their incentives to do so.

The main way that indigenous peoples in forested areas could get
access to significant amounts of capital for farming, logging, or other
productive activities would be to build alliances with nonindigenous
companies or individual producers, or with donors or nongovernmental organizations. The opportunities for such alliances have not been
sufficiently studied and vary widely depending on government policies
and local conditions. Another possibility is that outsiders will deforest
indigenous territories without the indigenous peoples' permission. In
many countries encroachment on indigenous territories by ranchers,
farmers, and drug traffickers, as well as forestry, mining, and energy
companies is a growing threat that affects areas where indigenous peoples already have formal land titles as well as areas where they do not.
These outside groups typically have much greater access to capital and
labor and are able to clear much larger areas of forest. Consequently,

addressing the invasion of indigenous territories is one of the most urgent tasks required to maintain low deforestation rates in these areas.

GOVERNMENTS HAVE KEPT PEOPLE FROM CLEARING FORESTS IN INDIGENOUS TERRITORIES

In a large percentage of indigenous territories in Latin America, government policies restrict forest clearing, by both indigenous peoples and others. Brazil alone accounts for approximately 40 percent of all legally recognized indigenous territories in the region. The Brazilian government considers these territories a type of protected area: indigenous peoples are allowed to clear forests for their subsistence use but not beyond that. In most other Latin American countries there is also a significant overlap between indigenous territories and protected areas. Hence one might argue that one reason that the indigenous territories have not suffered greater deforestation is that the governments have not allowed it. Indeed, many indigenous organizations have complained that the governments have restricted their activities in their traditional lands and taken decisions without consulting them.

As with the other hypotheses, this one probably has some validity, but it is definitely not the whole story. Indigenous leaders point out that they have conserved their forests since long before they were declared to be protected areas, and as we have seen above, in many places indigenous territories have achieved better conservation outcomes than strictly protected areas. There are also clearly many places where the indigenous peoples themselves have played a major role in protecting the forests in their territories, and not relied simply on governments. Governments *do* have a particularly important role to play in ensuring that indigenous territories are not illegally invaded by outsiders. This is particularly true in cases involving large territories with small indigenous populations who find themselves outnumbered and outgunned by outsider intruders. Brazil has done a reasonably good job of keeping outsiders from encroaching on legally recognized indigenous territories; however, most other countries have not.

Policy options for the future

Historically, the remoteness and poor soils of the large compact forest areas inhabited by indigenous peoples combined with their low population densities, high levels of poverty, and diverse, forest-dependent production systems were sufficient to conserve most of their forests. Over time, however, internal and external factors have increased the pressures on those forests. The indigenous communities have grown and have become more integrated into the market and more assimilated into Western culture. It has become harder for them to subsist by hunting, fishing, harvesting forest products, and cultivating small plots on riverbanks. New roads and markets, faster boats, and chain saws have made it easier for local communities to clear forests and take their products to market. Increasingly, we are witnessing the end of the hinterland. New technologies, better transportation infrastructure, and high commodity prices have made it practical and profitable to push into most of the areas that used to be too remote and unappealing.[16] This has led to an avalanche of nonindigenous farmers and companies moving into these areas.

If there is to be any hope of conserving a significant portion of these forests, the indigenous peoples must be given tools and support to protect their lands from outside encroachment. While the indigenous peoples themselves may also gradually decide to clear growing areas of forest, this threat pales in comparison with the current onslaught of outside interests. Most of the indigenous peoples have inhabited and maintained informal possession of their territories for generations and have clear legal rights to own them. Moreover, for the most part they have done so in a way that does not fundamentally destroy the resource base. They are some of the poorest groups in Latin America;

the sustainable management of their natural resources could provide a pathway out of poverty, but only if powerful interests do not dispossess and displace them.

Recent decades have seen a great deal of progress toward greater recognition of indigenous forest rights—symbolized by a large rise in the areas of indigenous territories that have received official government protection. However, in the last few years this trend appears to have slowed down, and some governments have become increasingly reluctant to take steps to stop the encroachment of the territories that they themselves have recognized. As these resources become more valuable, the pressure has grown to take them away from their original inhabitants.

International initiatives like REDD+ should provide strong impetus to strengthen indigenous peoples' rights to stop outside encroachment on their forests. Investing in the demarcation, titling, patrolling, governance, and economic development of indigenous territories offers a cost-effective option for reducing carbon emissions while defending human rights and improving the well-being of some of the region's most vulnerable groups. For the most part, however, this has yet to happen. While governments have spent hundreds of millions of dollars on forest conservation projects in Latin America, only a tiny percentage of that has gone to support indigenous peoples. If there is to be any hope of conserving a significant portion of the region's forest, that will have to change.

DAVID KAIMOWITZ *is the director of sustainable development at the Ford Foundation and a former director general of the Center for International Forestry Research (*CIFOR*).*

Wood engraving depicting a rainforest in New Grenada of the nineteenth century—today's Colombia.

7

RAINFOREST CONSERVATION

Impressive Results—Are They Enough?

A CENTURY AGO FEW people would have shared Charles Darwin's delight as he wrote in his diary on the HMS *Beagle* in 1836: "Among the scenes which are deeply impressed on my mind, none exceed in sublimity the primeval forests undefaced by the hand of man..." The dark and impenetrable rainforests inspired fear and were more of an obstacle than a vegetation cover worth preserving.

The first initiatives to protect tropical rainforests were not driven by a concern for their biodiversity, but by the desire to establish a "permanent forest estate" to preserve its timber value. In the late nineteenth and early twentieth centuries, colonial governments introduced the first forest management legislation, and they established forest departments, at least in those countries whose timber supplies were important to the European market. Other forests were protected in exposed zones to prevent soil erosion or as shelter belts against climatic effects. In Ghana the colonial government published a list of forest reserves in the mid-1920s, arousing an angry reaction from the local forest communities who lost their customary rights to plant and collect forest products like fruit, nuts, gum, fiber, and chewing sticks in these areas. By the 1930s the state forest reserves had reached almost 20 percent of the Ghanaian rainforest zone.[1] Most of these state forest reserves still exist, whereas the surrounding areas have practically all fallen victim to farming and are now riddled with unproductive farm bush. In countries where the forest service was less efficient, such as in neighboring Côte d'Ivoire, the state forest reserves (forêts classées) were far less resistant to deforestation by slash-and-burn farming, as can now easily be detected on satellite imagery (see plate 10).

Protecting rainforests: Not a priority until a few decades ago

EVEN AFTER WORLD War II, in the 1950s and 1960s very little attention was paid to protecting tropical rainforests for their biodiversity value—their extent was so immense that it was inconceivable that their biodiversity could ever be threatened. In his landmark study of the geography of deforestation, the author Michael Williams describes how, until the postwar economic boom, tropical forests were believed to be "exuberant, resilient and indestructible."[2] Little was known about their fauna and flora, unlike in the savannas and open forests of Africa and India where the wildlife could easily be observed—and hunted, which was already a serious concern for naturalists at the turn of the century. Worry about the abundance of animal species and scientific knowledge about them have always been functions of their observability.

The first protected areas in Africa, the Hluhluwe, Umfolozi, and St. Lucia reserves in South Africa, were established in 1895—long before conservationists could imagine that the unknown interiors of the large tropical rainforest blocks in Amazonia, the Congo Basin, or Borneo would ever be in danger. With the exception of the small-scale swidden agriculture of indigenous forest tribes, rainforest was only cleared on the fringes of these vast areas. Not even rough estimates of their extent existed and, in fact, it was not until the introduction of satellite remote sensing technology in the early 1990s that we could more or less accurately assess the enormity of these rainforest areas.

Remarkably, one of the scientists who was most knowledgeable about tropical rainforests in the postwar period, P. W. Richards, in his historically important book *The Tropical Rain Forest*[3] had already warned in 1952 about the consequences of continued rainforest destruction. It was an early hint of what was to become a much worse onslaught in the coming decades. It was only from the 1970s that priority was given to establishing protected areas of tropical rainforest. With some rare exceptions of tropical rainforest national parks having been established earlier (for example the Taman Negara National Park in Peninsular Malaysia in 1939, or the Salonga National Park in the Democratic Republic of the Congo in 1956), the majority of the protected areas in the rainforest biome were established after 1970. This includes some of the most famous ones: the Taï National Park in Côte d'Ivoire in 1972, the Gunung Leuser National Park in Sumatra in 1980, and the Jaú National Park in the Brazilian state of Amazonas, also in 1980. But these first rainforest national parks, which all still exist, were not more than samples of the biome and primarily of scientific interest. Nobody would have expected these few protected areas to be sufficient to represent the biome, nor did anyone see a need to establish a more extensive network of protected areas, as long as the threats were not obvious.

LARGE-SCALE DESTRUCTION CALLS FOR
PROTECTED FOREST AREAS

IN THE COURSE of the past forty-five years, since the fate of the tropical rainforests became a matter of widespread concern, the attention paid by the public, governments, and nongovernmental organizations has fluctuated greatly. Awareness took off with the publication of Adrian Sommer's article in the FAO's *Unasylva* journal[4] and became headline news in the 1980s with the publication of Norman Myers's landmark book, *The Sinking Ark,*[5] and his subsequent publications.[6] When the term "biodiversity" was coined by the scientists who published papers at the National Forum on BioDiversity, held in Washington in September 1986,[7] not only did the importance of biodiversity arrive in the minds of the informed public, it soon became clear that a very large part of the world's biodiversity depended on the future of the tropical rainforests.

Nongovernmental organizations played an important role in building awareness about the importance of tropical rainforests, the treasures they contained, and the threats that confronted them. WWF International made tropical forest conservation a main priority and launched many field projects in the 1980s. The Nature Conservancy, with support from the United States Congress, launched the Parks in Peril program in 1989, designed to protect twenty million hectares in Central and South America and the Caribbean. Conservation International was founded in 1987 specifically to conserve tropical biodiversity. Another organization from the United States, the Rainforest Alliance, was founded in the same year with a primary focus on the sustainable use of tropical forest products and its own certification scheme. The Rainforest Foundation founded by Sting and his wife Trudie Styler in 1989 focuses primarily on the protection of indigenous rights.

More radical nongovernmental organizations were also created to defend the cause of the rainforests and their inhabitants: Robin

Wood in Germany in 1982, the Rainforest Action Network in the United States in 1985, and the World Rainforest Movement in the United Kingdom in 1986. While most of these groups were founded in northern countries and began with a focus on the Amazon, they have since expanded their presence into southern countries. All these organizations have helped publicize the cause of the tropical forest in the 1980s. Newspapers and television channels followed with dramatic pictures of denuded tropical forest areas. The public in northern countries reacted with their emotions and also with their wallets to fund individual rainforest projects.

PUBLIC AWARENESS CONVERTED INTO INTERGOVERNMENTAL BUREAUCRACY

THE OUTCRY, ESSENTIALLY of the Western public and nongovernmental organizations, triggered a reaction in governmental and intergovernmental communities. Without much understanding and analysis of the root causes of deforestation, they created a fleet of agreements and policies to stop the destruction. The first of these was the Tropical Forestry Action Plan in 1985 (see box 2.1), followed by the creation of the International Tropical Timber Organization to promote sustainable forest management in the tropics. Then came the adoption of the United Nations Convention on Biological Diversity in Rio in 1992. The Earth Summit* also adopted the "Statement of Forest Principles," which eventually was followed by the United Nations Forum on Forests in 2000.

With the engagement of intergovernmental institutions, things gradually became quieter around the issue of rainforest destruction, partly because climate change had now become the number one global environmental concern. As the United Nations seemed to be taking care of the tropical rainforest problem, the media and the

---

* The UN Conference on Environment and Development (UNCED), also known as the Earth Summit, took place in Rio de Janeiro, Brazil, June 2–14, 1992.

broader public were inclined to believe the issue was in good hands. The concern over the destruction of tropical rainforests of the 1980s dissipated into debates at numerous international meetings that tied up much capacity in governments and nongovernmental organizations alike.

Eventually it became clear that the time for sounding the alarm (telling the world the rainforests were disappearing) and hoping for the best (waiting for the United Nations to save them) was past for the conservation community. New approaches in tropical rainforest conservation were badly needed. Deforestation was accelerating in South America and Southeast Asia, and with a limited number of protected areas scattered across the tropical rainforest biome, the future at the beginning of the 1990s looked bleak. Very bleak.

The boom years for protected rainforest areas

IN 1990 THE United Nations list of protected areas included 669 sites in the lowland tropical moist forest biome, covering some sixty-six million hectares. This corresponded to only a little more than 5 percent of the world's tropical rainforests and raised concern that this could be the end of progress toward the 10 percent target set at the World Parks Congress in Bali in 1982. Reaching 100 million hectares of protected rainforest areas in the twenty-first century seemed like an infinitely optimistic scenario.[8]

For a protected area to be recognized as a "forest protected area" on the United Nations list, it must fit into one of the six IUCN categories of protected area. This means that their primary purpose is preserving biodiversity and cultural values.* Forest areas that

---

* Primary objectives of IUCN Protected areas categories: Ia: science of wilderness protection; Ib: wilderness protection; II: ecosystem protection and recreation (national parks); III: conservation of specific natural features; IV: conservation through management intervention; V: landscape/seascape conservation or recreation; VI: sustainable use of natural resources.

are protected to ensure a clean drinking water supply, to avoid soil erosion, or as wind and fire breaks do not count as forest protected areas, nor do plantation forests, even if they fall within a larger forest protected area.[9] The Brazilian indigenous reserves that cover about 20 percent of the Brazilian Amazon were historically not listed as protected areas, even though they play an important role in Amazonian forest conservation.[10] To preserve the world's biodiversity, which to a very large extent hinges on the tropical rainforests, a far larger set of protected areas was needed, and the remaining forest outside protected areas also had to be managed sustainably.

TABLE 7.1. **Percentage of tropical rainforest area under protection**

Humid tropical forest types according to the UNEP/WCMC Global Forest Map (GFM) that are protected under IUCN categories I through IV and I through VI. Not included are some areas of the map that were unresolved. Source: Schmitt et al. (2008).[11]

GFM FOREST TYPE	FOREST AREA IN MILLION HA	% PROTECTED IN IUCN CATEGORIES I–IV	% PROTECTED IN IUCN CATEGORIES I–VI
Trop. lowland evergreen broadleaf rainforest	648.9	10.3	20.8
Trop. semi-evergreen moist broadleaf forest	84.3	17.7	26.4
Trop. freshwater swamp forest	53.6	7.2	11.3
Trop. lower montane forest	44.8	12.7	17.5
Trop. upper montane forest	47.6	18.2	26.1
Trop. mangrove forest	13.7	13.1	20.6
TOTAL	892.9	11.4	20.9

Despite the gloomy prospects, a remarkable thing happened: rather than slowing, the establishment of protected rainforest areas actually accelerated. It was as if the rainforest body had developed an immune reaction to the deforestation virus. By 2003 the number of protected areas had grown more than fourfold and now covered 245 million hectares! Not 10 percent but 23 percent of the rainforest was protected according to the United Nations list of protected areas.[12] Why this sudden motivation in tropical countries to put rainforest under protection—and was the protection real? A number of these newly declared protected areas had not been registered under an IUCN category. Discounting those, the protected area coverage was 20.9 percent, and if only the areas belonging to the strictest protection status of IUCN categories I through IV were counted, the coverage dropped to 11.4 percent.[13] But even then, the area of tropical rainforests under strict protection had more than doubled, and tropical rainforests were now better protected than the world average of all forest types (see table 7.1).

WEAK SIGNALS FROM RIO CALL FOR ALTERNATIVE ACTION

WHAT HAD HAPPENED in these thirteen years? Was this a result of the United Nations Conference on Environment and Development in Rio of 1992 and its non–legally binding forest principles? The forest principles fell far short of the strategic goals published by the IUCN, United Nations Environment Programme, and WWF the year before, in the second version of the World Conservation Strategy.[14] Or did the new Convention on Biological Diversity (CBD) signed in Rio by 168 countries unleash a wave of new protected areas?

The CBD required its signatories to develop strategies and action plans, but by 2002 fewer than half of them had established strategies. The CBD was hampered in its first decade by a lack of clear targets because biodiversity is difficult to monitor. Institutional bureaucracy within the CBD also kept the convention from playing a significant role in its first decade. A program of work on protected

areas was only adopted more than ten years later, at the CBD Conference of the Parties in Kuala Lumpur in 2004. It is therefore unlikely
that the sudden bloom of protected areas in rainforest countries of
the 1990s had much to do with Rio and its multilateral agreements.

THE POWER OF MEASURABLE TARGETS

IN FACT, THE hope that multilateral agreements could save the
tropical rainforests and their biodiversity started waning soon after
the Rio conference. The slow progress of the United Nations system, which all too often descends to the level of the lowest common
denominator when agreeing on a standard, left nongovernmental organizations and the more progressive national governments
frustrated. Meanwhile, information on some of the highest-ever
deforestation rates in tropical areas became public.

Disillusioned with the intergovernmental process, some of the
larger nongovernmental organizations began to look for other ways
to act. Why not establish targets independently, publicize them
through global conservation networks, and promote them with
campaigning methods? Among these new targets was a goal set at
the World Parks Congress in Bali in 1982 that "protected areas cover
at least 10 percent of each biome by the year 2000." WWF adopted
this target after the Rio conference and extended it to mean "at
least 10 percent protected areas in each major forest type by the
year 2000." They combined it with another target to increase the
productive forest areas under Forest Stewardship Council (FSC)
certification. The FSC had been founded the year after the Rio conference in 1993 as a way to harness market forces and establish
standards for sustainable forest management where some of the
intergovernmental efforts had failed.

There is something mystical about targets—they attract attention from sympathizers and skeptics alike. WWF forest targets
triggered a reaction from an unexpected source—the president

of the World Bank, James Wolfensohn. In 1998, the World Bank, which had been criticized for its forest policy, supported WWF's targets and proposed an alliance to set further targets to establish new protected forest areas, effectively manage existing ones, and certify production forests. The World Bank–WWF Alliance for Forest Conservation and Sustainable Use contributed substantially to protected rainforest areas coverage and played a crucial role in the success of the Amazon Region Protected Areas plan (see box 9.2 and Cláudio Maretti's Specialist's View on p. 204).

THE NEED FOR INSTITUTIONAL ENGAGEMENT

THE 10 PERCENT target for protected forest areas has often been criticized—mostly because it was misunderstood as a signal that 10 percent would be enough to safeguard biodiversity. Even serious conservationists demanded to know what should happen to the other 90 percent, whereas some university scholars, on the other hand, felt that 10 percent was either unrealistic or an arbitrary figure with little meaning.[15] Still others worried that such targets would only lead to "paper parks"—areas that are protected on a map but do not have effective management and enforcement.

Such criticism is not without justification. We have heard it again about the Convention on Biological Diversity's target of at least 17 percent of terrestrial and inland water areas being conserved as well-managed protected areas by 2020.[16] But whatever people thought about the 10 percent target for forests, it worked! Of course 10 percent protected forest cover is not enough, but it is twice as much as 5 percent, and such arguments count when decision makers have to be convinced. Politicians tend to like and understand targets. And as long as it is made clear that a target should be understood as a time-bound objective, a milestone to achieve the ultimate goal—protecting an ecosystem of huge global importance—there is nothing wrong with using targets as a tool. But there is another aspect to this as well: when targets come across like homework assigned by a schoolteacher, the results are predictable.

Obviously, targets alone do not do the trick. They are only effective if the accompanying conditions are conducive to change: public awareness of the threats to ecosystems; nongovernmental organization campaigning and publicity with statistical and photographic evidence on the drivers of deforestation and the related fate of forest-dependent people; financial support and favorable working relationships between governments and nongovernmental organizations; and most importantly, political will and determination (see box 7.1).

Regional agreements may also contribute importantly to conservation forest conservation goals. The Yaoundé Declaration resulted from a meeting of Central African heads of state in Yaoundé, Cameroon, in 1999. It led to the Central African Forest Commission (COMIFAC) and fostered the creation of a large cluster of protected areas in Cameroon, the Republic of the Congo, the Central African Republic, and Gabon (see box 7.2). The Yaoundé Declaration helped address the historical underrepresention of protected areas in West and Central Africa compared to other tropical rainforest regions: by 2003 they made up 14 percent of the moist tropical forest area, half the percentage reached in South America.[17]

BOX 7.1

The role of leadership:
An encounter with President Cardoso

It was sheer luck that I managed to catch the last flight from Manaus to Brasília on that November day in 1998. I had been visiting a community development project in Silves, some 300 kilometers east of Manaus, together with two of my colleagues in the WWF: Twig Johnson and Bob Buschbacher. On our way back through the extensive várzea lakes, inundated lowland forest areas that connect to the Amazon River, our boatman lost his way and got stranded in some shallows with the propeller of the outboard motor stuck in the mud. It was an exceptionally dry year in the Amazon, and water levels were low. The smoke from forest clearing near Manaus cast a haze that hid the sun—and the shore. All we could see were the eyes of some caimans protruding from the muddy waters.

What an embarrassing situation—between us, we had decades of experience in tropical forests, but here we were, stuck in the mud in the middle of this várzea lake, equipped with one useless paddle. We had no idea where we were, and the boatman was not much help. But the real problem was that I might miss an important appointment with the president of Brazil in Brasília early the next morning! As we could not move the boat while sitting in it, we had no choice but to take our trousers off and plow through the almost hip-deep mud, pushing the boat in the hope of finding some deeper water. It took six hours to liberate ourselves from this silly situation and fix the outboard motor. We arrived at the Manaus airport in the late evening, still splashed with mud, but more than happy to find out there was a late plane leaving for Brasília.

President Fernando Henrique Cardoso addressed me in his perfect French and chuckled when he heard about our fate in the várzea the previous day. He knew that I had not just come to pay him a courtesy visit, nor was he shocked to hear that the WWF had a target to increase the protected areas coverage to 10 percent in all forest types and to promote independent certification for forestry operations. He had witnessed the onslaught on the Amazonian forests during his presidency, and it did not leave him indifferent. But Cardoso also knew that he needed public and political support for such an ambitious plan or it would not be feasible. A few months later the World Bank president, James Wolfensohn, announced support for the same targets. The partnership became known as the World Bank–WWF Alliance for Forest Conservation and Sustainable Use, and was followed by the financial commitment from the chairman of the Global Environmental Facility, Mohamed El-Ashry, and the German development bank KfW.

It took four years of intensive planning and fundraising for a trust fund until an agreement for the Amazon Region Protected Areas Program came to fruition. It was signed by President Cardoso and the president of the Brazilian parliament, the World Bank, the GEF, and the WWF at the World Summit on Sustainable Development in Johannesburg in 2002 and launched the following year. And it became the world's largest tropical rainforest conservation program ever, well on target to secure fifty million hectares of additional protected areas in the Brazilian Amazon—an area the size of Spain (see Cláudio Maretti's Specialist's View on p. 204). The vision and determination shown by President Cardoso makes my encounter with him, the day after the embarrassment in that muddy várzea lake, one of the most memorable events in my conservation career. It is strong evidence that political will and leadership, those rare commodities, can make all the difference when it comes to rescuing the future of the world's tropical rainforests.

BOX 7.2   **The Yaoundé Process**

In 1999, the Central African heads of state and the Duke of Edinburgh, then the president of WWF International, met in Yaoundé, Cameroon. The result was the Yaoundé Declaration, which in turn led to the creation of the Central African Forest Commission (COMIFAC). The Yaoundé Process contributed to establishing new protected areas in southern Cameroon: Lac Lobéké, Boumba Bek, and Nki National Parks, as well as the large complex of protected areas between Cameroon, the northern part of the Republic of the Congo, and the southern tip of the Central African Republic, which is now supported by the Tri-National de la Sangha Trust Fund. The Yaoundé Process also resulted in thirteen new protected rainforest areas in Gabon in 2002, declared by President Omar Bongo in the advent of the World Summit on Sustainable Development (see plate 4).

Such declarations may be motivated by the desire for international visibility and may not be followed by the necessary management support, but they still send the right signal, not least to the international community, which can organize support. At the World Summit in Johannesburg the US Secretary of State, Colin Powell, made a pledge of USD 53 million and in fact launched the Congo Basin Forest Partnership, a facilitating platform that brings together some seventy partners, including African countries, donor agencies, international organizations, nongovernmental organizations, scientific institutions, and the private sector.[18]

CONSERVATION FATIGUE

IN THE LAST ten years the growth of protected rainforest area and indigenous reserves has slowed considerably. There are a number of possible reasons for this:

1) The demand for forest land that can be converted to agricultural land or used for logging or mining remains strong and may increase, especially in Africa.

2) Protected tropical rainforest areas may not generate any cash revenue, and their ecosystem services do not appear on national balance sheets, so governments do not recognize their value.

3) Some of the organizations that campaigned for protected areas and were involved in partnership processes such as the Amazon Region Protected Areas plan develop "rainforest fatigue" from the strain of finding sustainable funding for these areas, and campaigns for other conservation goals also become important.

These factors working against conservation measures raise the question of whether primary forests outside existing reserves have any chance of surviving the next decades.

BOX 7.3   **The Heart of Borneo initiative**

RODNEY TAYLOR

Background

The Heart of Borneo is an area of over twenty-three million hectares, straddling three countries: Brunei, Indonesia, and Malaysia (see plate 12). In 2007, the governments of these countries signed a declaration[19] that committed them to cooperate to manage and conserve the forest resources within the Heart of Borneo through a network of protected areas, productive forests, and other sustainable land uses. The goal was "maintaining Bornean natural heritage for present and future generations."

Though still largely unexplored, the Heart of Borneo has a staggering number of unique species found nowhere else in the world. One of Asia's last great expanses of tropical forest, the area is home to 6 percent of the world's biodiversity, from the orangutan to the world's largest flower, and it contains the headwaters of fourteen of Borneo's twenty major life-sustaining river systems. The Heart of Borneo is also the cultural epicenter of indigenous highland communities.

Threats to Borneo's forests

A century ago, most of Borneo was covered in forest. The island has since undergone a massive transformation as forests were cleared, converted to other land uses, or degraded. The rate of loss and degradation has accelerated in the last two decades. In 2005 the WWF[20] projected a forest loss of 17.28 million hectares from 2000 to 2020, meaning that only 23 percent of Borneo would remain forested (see plate 11). A study in 2013 reinforced this analysis, predicting that if current deforestation rates continue, 21.46 million hectares will be lost from 2007 to 2020.[21]

The underlying cause of forest loss and degradation in Borneo is weak governance that drives short-term thinking. Pressures are amplified by in-migration, road development (95 percent of deforestation in Borneo occurs within five kilometers of a road[22]), and the high profitability of activities such as oil palm plantations and coal mining. Fires are also a major threat. Fires are used to clear land, but they often spread to burn out of control on drained peatlands (an area of around one million hectares was drained for Indonesia's failed rice mega-project) and in forests made drier by El Niño events or large canopy gaps resulting from poor logging practices. Logged and abandoned timber concessions are also vulnerable to encroachment and conversion.

In stark contrast to the rest of the island, the forests of the Heart of Borneo remain relatively intact. A recent study by the WWF[23] estimated the remaining forest types as a percentage of their historical extent. Particularly the upland rainforests and the montane forest remained relatively well preserved with 82 and 89 percent.

However, the forests of the Heart of Borneo are coming under increasing pressure. Lower-altitude natural forest is being converted to oil palm plantations and other agricultural crops. Illegal or unsustainable logging, forest fires, mining, and overhunting of wildlife are also major threats. An increase in climatic variability will exacerbate these threats.

Conservation targets and strategies

PROTECTED AREAS

Protected areas are a key component of strategies to conserve the Heart of Borneo. An extensive, representative, effective, and well-connected protected area system is critical to support biodiversity and buffer the area against climate change. The Heart of Borneo needs new protected areas to improve connectivity and make sure all forest types are adequately represented in the protected area system (see table below). A comprehensive survey of the management effectiveness of the current protected areas is also a high priority. Protecting lowland forest is vital, given how little remains in other parts of Borneo. This is the prime habitat of the pygmy elephant, orangutan, and rhinoceros—some of the most endangered species in the world.

Heart of Borneo protected areas
Source: S. Wulffraat (2012).[24]

ECOSYSTEM	AREA PROTECTED (HECTARES)	% PROTECTED
Lowland rainforest	335,500	9.6
Upland rainforest	1,317,700	18
Montane forest	995,900	29
Heath forest	<30,000	<1
Limestone forest	<20,000	<1

LOGGING
Well-managed logging concessions are another key conservation strategy. In Sabah, for example, the Deramakot and Ulu Segama-Malua Forest Reserves are managed for sustainable wood production and feature programs to restore forests in damaged areas. Together they harbor the largest orangutan population in Malaysia. Such forest stewardship, motivated partly by a commercial interest in maintaining wood supply, helps protect natural forests from illegal logging, encroachment, or conversion to other land uses.

PALM OIL WITHOUT FOREST LOSS
Producing palm oil without clearing more forest is another key element of conservation. Achieving this will require the reform of policies governing land use and permit allocation, and actions by the private sector

to create deforestation-free supply chains and improve plantation practices in compliance with voluntary sustainability standards. For example, the Roundtable on Sustainable Palm Oil forbids expansion into primary forests, areas with high conservation values, and areas where the holders of legal or customary rights have not given their free, prior, and informed consent. In Indonesia vast areas of degraded land can be traced back to palm oil permit holders who cleared forests with fraudulent intent to sell the timber and abandon the land, or whose planned plantations failed due to lack of capital, labor, or know-how.

FINANCING

The Heart of Borneo initiative needs innovating financing to fund the activities in its implementation plans and to help intact forests compete with land uses that generate higher short-term revenues. There are several such mechanisms in development:

- REDD+ (see Chris Elliott's Specialist's View on p. 235): Central Kalimantan is a high-priority area for REDD+ investment in Indonesia. In 2011 the province produced a low-carbon development plan. Fully implementing this plan will require many institutional, regulatory, and policy reforms, but it has shown the potential to reduce forest loss and degradation and to attract significant donor support and REDD+ finance.
- To protect watersheds in the Heart of Borneo area, the WWF and the nongovernmental development agencies CARE and the International Institute for Environment and Development (IIED) are supporting a project where downstream users of river water pay upstream communities for their protection of forests that filter water entering the river system.
- The Brunei government is cooperating with the Japanese National Institute of Technology and Evaluation to explore the potential to generate revenue from licensing research on prospective commercial uses of the rich genetic resources of the rainforest.

- Malaysia's Malua BioBank: The Sabah government has licensed conservation rights for fifty years to the Malua BioBank, and a private investor has committed up to USD 10 million for the rehabilitation of the Malua forest reserve.

Future challenges

The key challenge for the governments working together on the Heart of Borneo initiative is to integrate the full value of forest ecosystem services into their development plans and to enable land use choices that take into account the risks to business and society if those services decline. This will require robust policy reforms, new incentives, and effective enforcement. The success of the Heart of Borneo initiative will also depend on continued support from civil society organizations, the private sector, and donor organizations.

RODNEY TAYLOR *is the director of* WWF *International's forest program.*

Primary forests: Important but dwindling

PRIMARY TROPICAL RAINFOREST is defined by the FAO as "naturally regenerated forest of native species, where there are no clearly
visible indications of human activities and the ecological processes
are not significantly disturbed." Because this forest has a much
higher biodiversity value than degraded forests, it is essential to
assess how much primary forest remains, and at what rate these forests are shrinking.

Unfortunately we have only the most approximate information.
Some tropical countries do not make a distinction between primary and secondary forests in their reports to the Forest Resources
Assessment of the FAO, or simply report the forest area included
in protected areas. What about high-resolution remote sensing?
Selective logging can be detected by its forest roads and timber land-
ings, but as soon as secondary vegetation starts covering these scars,
logged areas are hardly discernible on satellite imagery. This is particularly true in areas that have been exploited with low-impact
logging practices. To reliably distinguish primary forests from secondary forests it would be necessary to compare the forest cover
with maps of logging concessions, but this would still not include
illegal logging outside approved concessions.

Of the tropical rainforest countries that reported data to the For-
est Resources Assessment (FRA 2010) of the FAO, those with the
largest areas of primary forest are in South America: Brazil, Peru,
Bolivia, and the Guianas. In Southeast Asia, Indonesia and Papua
New Guinea reported the largest areas of primary forest, and in
Africa it was Gabon (see table 7.2). Unfortunately neither the largest
African rainforest country, the Democratic Republic of the Congo,
nor Cameroon reported any data on their primary forest cover. The
Democratic Republic of the Congo in particular is expected to still
harbor large areas of primary forests. The accumulated global and

regional statistics on primary forests in FRA 2010 are therefore so deficient that they are hardly worth citing.

TABLE 7.2. Primary forest area trends

Primary forest area trends in selected rainforest countries between 1990 and 2010 in millions of hectares (extracted from the FRA 2010 country tables of the FAO).[25] Some countries did not report information on the area of primary forest, including some of the largest rainforest countries such as the Democratic Republic of the Congo or Cameroon. Others used the current area of forest in protected areas as a measure of their primary forest area. Some of the loss of primary forest areas was attributable to conversion for agricultural purposes, but most of the lost primary forest was reclassified into "other naturally regenerated forest" because of various forms of degradation, including logging and other interventions.

COUNTRY	PRIMARY FOREST AREAS (MILLION HECTARES)				AS % OF ALL NAT. REGENERATED FORESTS IN 2010
	1990	2000	2005	2010	
Brazil	530.0	501.9	488.3	476.6	93.1
Peru	62.9	62.2	61.1	60.2	90.2
Indonesia	no data	49.3	47.8	47.2	52.0
Bolivia	40.8	39.0	38.2	37.2	65.0
PNG	31.3	29.5	28.3	26.2	91.5
Gabon	20.9	17.6	16.0	14.3	65.2
Suriname	14.2	14.1	14.1	14.0	95.0
Colombia	8.8	8.7	8.6	8.5	14.2
French Guiana	8.0	7.8	7.7	7.7	95.2
Rep. of the Congo	7.6	7.5	7.5	7.4	33.3
Guyana	no data	6.8	6.8	6.8	44.7
Malaysia	3.8	3.8	3.8	3.8	20.5
Guatemala	2.4	2.1	2.0	1.6	46.5
TOTAL OF THESE 13 COUNTRIES	—	750.3	730.2	711.5	
Annual red. in %		0.54%		0.51%	

However, some conclusions can be drawn at the level of individual countries. Certain countries seem to be losing primary forests faster than they are losing forest cover overall, which suggests rapid opening up of primary forests and the expansion of selective logging. Some of the heaviest losses in the past fifteen years have been in Gabon and Papua New Guinea (see box 4.2). Agricultural expansion at the cost of primary forests is still common in South American countries, as happened on a large scale in the Brazilian arc of deforestation for soybean and cattle pastures until 2004. In absolute terms, Brazil still led the score of primary forest decrease with about twelve million hectares lost in the 2005–2010 period. Only a part of this loss is attributable to selective logging. The countries of the Guiana Shield that maintain relatively large areas of primary forest (Guyana, Suriname, and French Guiana), on the other hand, also show some of the lowest percentages of annual primary forest loss.

Tropical rainforest conservation strategies urgently needed

SETTING TROPICAL RAINFOREST areas aside for the purpose of preserving biodiversity and protecting the rights of indigenous people is a relatively recent phenomenon. It has only become a priority since commercial pressure for land conversion became intense, mainly in South America and Southeast Asia. As agricultural commodity markets become more global, these pressures can go nowhere but up, which means protecting tropical rainforest areas must remain a high priority. Certain regions and countries still do not have enough protected areas to preserve even a fraction of their biodiversity. Increasing fragmentation of the remaining forest blocks, and the fringe effects that come with it, expose tropical rainforests to what some researchers call a "double cocktail": rapid climate change and greatly reduced range sizes for many species.[26]

The vast expanses of primary tropical rainforests that were once believed to guarantee the preservation of biodiversity are dwindling rapidly. They are lost either by transformation to agricultural land or degradation by selective logging and infrastructure development.

It is not inconceivable that in a few decades no large forest areas will remain untouched by commercial exploitation outside protected areas and indigenous reserves. Although selective logging has had less severe consequences than anticipated,[27] there is no substitute for sufficiently large areas of primary rainforests for biodiversity conservation. Therefore, the maintenance, enforcement, and expansion of the current protected areas system in tropical rainforest areas is still essential for the conservation of the world's biodiversity. It follows that we will need to restrict the expansion of selective logging into hitherto untouched primary forests. The Forest Stewardship Council has developed a designation of "high conservation value forests" to protect particularly valuable primary forest areas, and the complementary concept of "intact forest landscapes" has been put forward by a group of nongovernmental organizations including Greenpeace and the World Resources Institute as a tool to identify future protected areas (see box 7.4).

Protected areas are sometimes portrayed as a playground for narrowly focused scientists and somewhat unworldly naturalists. Another criticism, with some justification, is that protected areas deprive local communities of their indigenous rights. But it is a mistake to disregard the ecological and economic services that protected forest areas provide. According to the analysis of the TEEB report, a sixth of the world's population depends on land in protected areas for a significant part of their livelihood. And this comes in addition to the globally important roles of sequestering carbon, protecting water cycles, and fostering climate stability. The real problem again is that their ecosystem services are not accounted for in national statistics, and therefore often count less than the value of a few tons of beef produced on an equivalent area.

1 Distribution of the "tropical and subtropical broadleaf forest" biomes

The moist broadleaf forests (dark green) represent the original distribution of tropical rainforests. Some areas are not tropical rainforests in the strict sense, for example, the moist deciduous (monsoon) forests of central and northern India, or the subtropical evergreen forests of southeastern China. This map is based on the definition of ecoregions,[i] and was produced by WWF US.

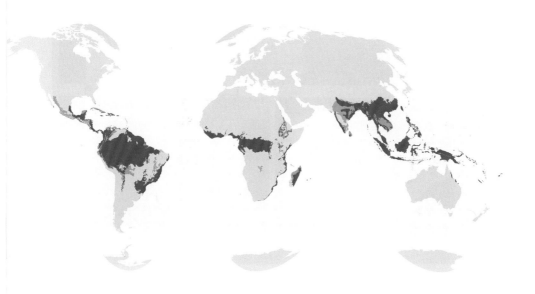

■ Tropical and subtropical moist broadleaf forests
▨ Tropical and subtropical dry broadleaf forests

[i] D. M. Olson, E. Dinerstein, E. D. Wikramanayake, N. D. Burgess, G. V. N. Powell, E. C. Underwood, J. A. D'Amico, I. Itoua, H. E. Strand, J. C. Morrison, C. J. Loucks, T. F. Allnutt, T. H. Ricketts, Y. Kura, J. F. Lamoreux, W. W. Wettengel, P. Hedao, and K.R. Kassem, "Terrestrial Ecoregions of the World: A New Map of Life on Earth," *Bioscience* 51, no. 11 (2001): 933–38.

4 Forest canopy in Ogooué-Lolo Province, Eastern Gabon

This rainforest area lies in a transition zone between the "moist evergreen" and the "moist semi-deciduous" type. Some emergent trees are seasonally leafless. This area is part of an FSC-certified timber concession of the Precious Woods company. A selective timber cut had taken place before this photo was taken. (Photo by Max Hurdebourcq.)

5 The kapok tree or silk cotton tree (*Ceiba pentandra*)

This species is unique because it occurs in Central and South America as well as in African rainforest and savanna areas, unlike practically all other rainforest trees, which have much more restricted distribution areas. Its seeds are wind-dispersed but seed pods may float across the sea. The tree now also occurs in South and Southeast Asia. Some of the largest kapok trees can be found in African rainforests (as in this picture) where they reach heights of up to sixty meters or more and diameters of up to three meters above the massive buttress roots. (Photo by Max Hurdebourcq.)

EXTRACTS FROM THE HIGH-RESOLUTION GLOBAL FOREST COVER MAP

Source: Hansen/UMD/Google/USGS/NASA.

M. C. Hansen, P. V. Potapov, R. Moore, M. Hancher, S. A. Turubanova, A. Tyukavina, D. Thau, S. V. Stehman, S. J. Goetz, T. R. Loveland, A. Kommareddy, A. Egorov, L. Chini, C. O. Justice, and J. R. G. Townshend, "High-Resolution Global Maps of 21st-Century Forest Cover Change," *Science* 342 (2013): 850–853, doi:10.1126/science.1244693.

- ■ Forest Loss 2000–2012
- ■ Forest Gain 2000–2012
- ■ Both Loss and Gain
- ▨ Forest Extent

6 Deforestation by shifting cultivation in the Democratic Republic of the Congo

Small-scale shifting cultivation spreading in wide bands along the roads leading north from Lisala and Bumba on the Congo River in the Democratic Republic of the Congo. Abandoned farms show forest gain north of Bumba (blue areas).

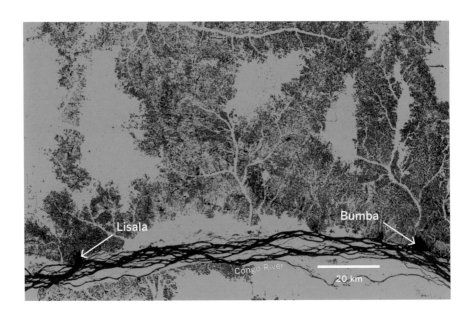

Forest fragmentation due to oil palm and timber plantations in Sumatra

The Riau Province of Sumatra, Indonesia, experienced particularly heavy deforestation and forest conversion in the 2000–2012 period. The use of fire for forest clearing has caused severe air pollution in Singapore. Pink areas are forests replaced by tree crops. Deforestation did not even spare Tesso Nilo National Park, a forest area with particularly high biodiversity.

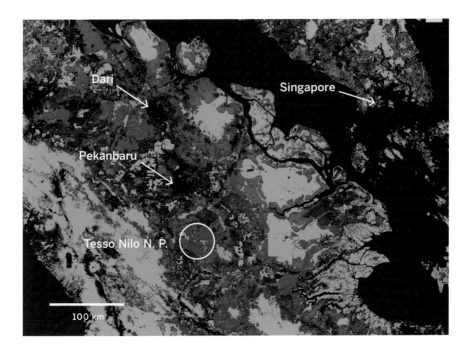

8 Deforestation hot spot in the Brazilian state of Rondônia

This satellite image shows the typical "fishbone" pattern of soy and cattle farms in the southern arc of deforestation of the Brazilian Amazon. Farms that were cleared in southeast Ariquemes before 2000 appear as black areas. The Pacaás Novos National Park contains the deforestation front effectively.

Deforestation along the Brazilian highway BR-163 between 2000 and 2012

Large commercial farms spread up to one hundred kilometers on both sides of the "soy highway" BR-163 between Cuiabá and Santarém. Farther north, smallholder farming is predominant.

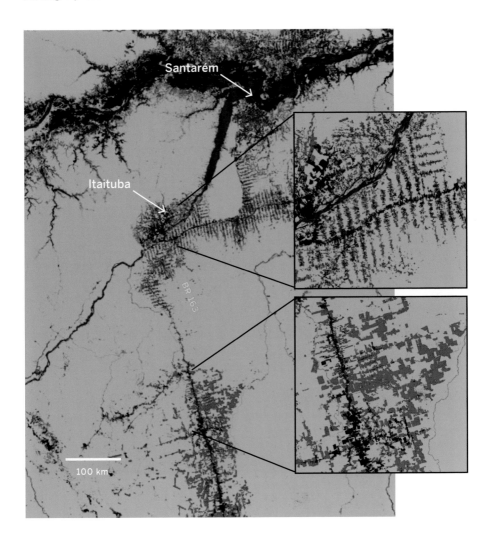

10 Forest reserves in western and central Ghana

Most forest reserves (light green areas) were established before World War II. A few of them have been converted to national parks, but the majority contain forests that have been selectively logged. The fact that these forest reserves still exist as forest, in contrast with the reserved forests in Côte d'Ivoire, for example, can be considered a success. However, the surrounding areas have been cleared, which exposes these reserves to the negative effects of fragmentation, including climate change and particularly (illegal) overhunting.

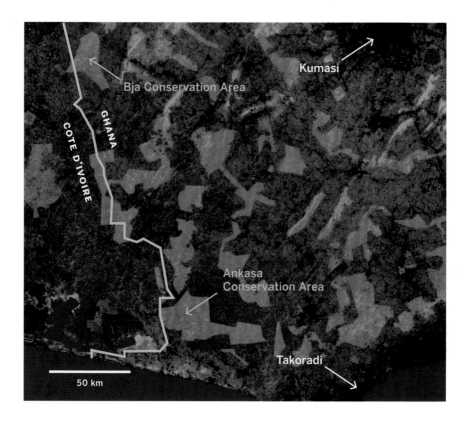

11 Deforesting of Borneo since 1973

Borneo has an area of 737,000 square kilometers—about the size of Germany. Three-quarters were still under intact forest in 1973. From 1973 to 2010 about 30 percent of the forest area was lost. By 2010 fewer than half the remaining forest areas were intact. From Gaveau et al. (2014).[iii]

1973 FOREST COVER

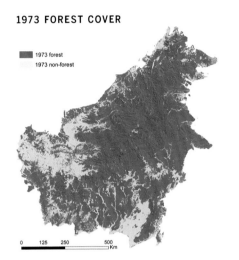

- 1973 forest
- 1973 non-forest

0   125   250   500
                             Km

FOREST COVER CLEARANCE

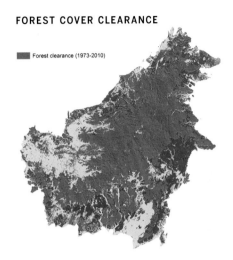

- Forest clearance (1973-2010)

2010 INTACT, LOGGED FOREST AND PLANTATIONS

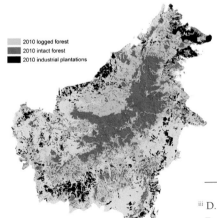

- 2010 logged forest
- 2010 intact forest
- 2010 industrial plantations

[iii] D. L. A. Gaveau, S. Sloan, E. Molidena, H. Yaen, D. Sheil, N. K. Abram, M. Ancrenaz, R. Nasi, M. Quinones, N. Wielaard, E. Meijaard, "Four Decades of Forest Persistence, Clearance and Logging on Borneo," PLOS ONE 9, no. 7 (2014): e101654, doi:10.1371/journal.pone.0101654.

The "Heart of Borneo" conservation effort

Forest cover of Borneo in 2012 with "Heart of Borneo" boundaries and protected areas. Source: map provided by WWF Indonesia (S. Wulffraat, K. Fahmi Faisal, and I. B. Ketut Wedastra).

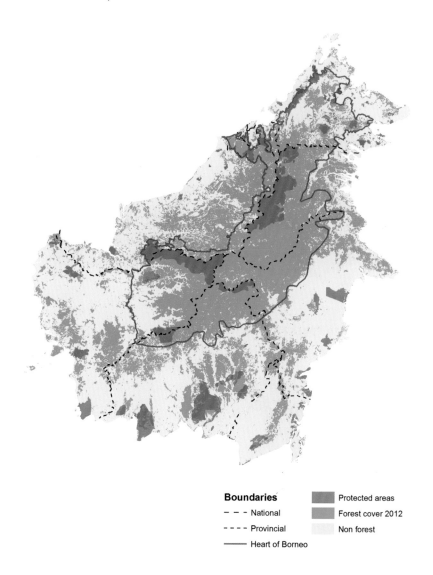

Boundaries

- – – National
- - - - Provincial
- ——— Heart of Borneo

■ Protected areas
■ Forest cover 2012
□ Non forest

13 Oil palm plantation and mill of Lahad Datu, Sabah (Malaysia)

Photo by WWF Malaysia/Mazidi Abd Ghani.

14 Protected areas and indigenous territories in the Amazon biome in 2010

Map updated 2013. Copyright WWF Living Amazon Initiative.

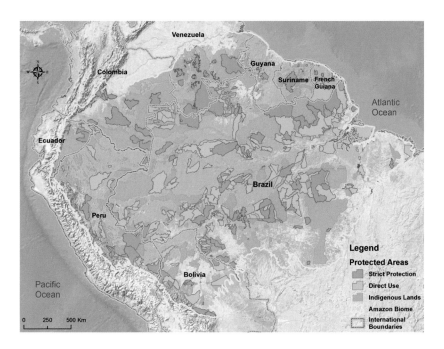

Rainfall anomalies and water deficit in the Amazon during drought years 2005 and 2010

A and *B*: satellite-derived standarized anomalies for dry-season rainfall for the droughts of 2005 and 2010 in the Amazon Basin. *C* and *D*: the difference between the maximum climatological water deficit (MCWD) in a twelve-month period (October to September) and the ten-year average for the years 2005 and 2010. This is a measure of drought intensity that correlates with tree mortality. Reproduced with permission from Lewis et al. (2011).[iv]

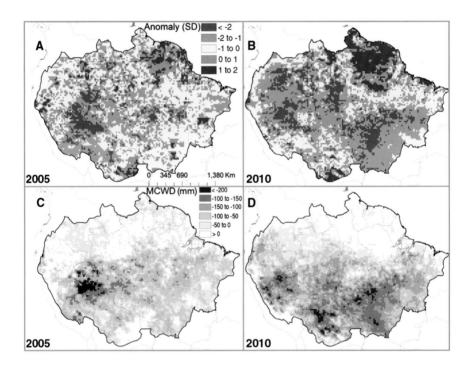

[iv] S. L. Lewis, P. M. Brando, O. L. Phillips, G. M. F. van der Heijden, and D. Nepstad, "The 2010 Amazon Drought," *Science* 331 (2011): 554, doi:10.1126/science.1200807.

16 Uncontacted tribe photographed by the Brazilian National Indian Foundation (FUNAI)

This photo, taken with a powerful zoom lens near the border with Peru, in the state of Acre, shows men painted in war colors, pointing their arrows at the plane. Proving the existence of uncontacted tribes is a requirement for their protection. Many of them are threatened by illegal loggers and drug traffickers.
Photo copyright Gleilson Miranda/FUNAI.

17 Rio Juruena, a tributary of Rio Tapajós, forming the border between the Brazilian states of Mato Grosso and Amazonas, 2014

Copyright: Zig Koch/WWF Brazil.

18 Montanhas do Tumucumaque National Park

A monadnock (inselberg) in the Montanhas do Tumucumaque National Park in the Brazilian state of Amapá, a virtually unexplored rainforest area the size of the Netherlands, bordering French Guiana. Copyright Zig Koch/WWF Brazil.

BOX 7.4   **Tools to prevent primary forest degradation**

Two complementary tools have been developed to protect large
expanses of primary forests, or at least contain their degradation.
These tools have been designed to preserve forest biodiversity, to
reduce carbon emissions from forest loss and degradation, and to
promote sustainable forest management practices.

High conservation value forests

The concept of high conservation value forests was first introduced by
the FSC in 1999 as part of their approach for certifying timber companies. The concept is used to identify forest areas that are particularly
valuable for biodiversity or local people. Management options for these
areas are designed to preserve or enhance their ecological and socioeconomic value.

Intact forest landscapes

The concept of an intact forest landscape is a relatively new approach
for identifying and mapping large areas of primary forest. They are
defined as an unbroken expanse of natural ecosystems, without significant signs of human activity, that make up an area of at least
500 square kilometers. About 600 million hectares of intact forest
landscapes are still found in the dense tropical and subtropical forest
biomes, primarily in Brazil and Central Africa. A group of researchers
supported by Greenpeace has created a global intact forest landscape
map based on high spatial resolution satellite imagery.[28] It has been
designed to help plan future forest protected areas and to stimulate
sustainable forest management practices.

PROTECTED AREAS ALONE WILL NOT BE SUFFICIENT

TROPICAL RAINFOREST CONSERVATION can never rely entirely on protected forest areas. It would be a serious mistake to disregard the importance of secondary forests—for one thing, they have potential to regenerate to a near-primary condition. Some of the forest transition cases described in chapter 5 have shown that such areas can maintain an important part of the original species diversity. Some timber concession areas in the Congo Basin have been found to harbor higher densities of certain species than adjacent national parks.[29]

An analysis of 292 protected areas in the Brazilian Amazon found that no specific governance regime and no specific IUCN category of protected area provides a sufficient guarantee for protection. Remote protected areas with less population pressure and less infrastructure, such as roads, are less prone to deforestation, irrespective of the protective status of the reserve. While such an effect is understandable, the more interesting finding was that the Brazilian indigenous reserves provided at least as much protection as strictly protected areas when deforestation pressure was high in surrounding areas.[30]

The conservation value of secondary tropical rainforests, which may be degraded to various degrees, should therefore be recognized in balance with efforts to establish protected areas and safeguard primary forest areas. Conservation strategies should include protection and sustainable management together, and they should enhance the role of indigenous forest communities. The overall aim of conservation strategies should not simply be to maximize protected area coverage—as important as protected areas are—but to make sure the erosive process of agricultural conversion and logging expansion into primary forests is mitigated. This will mean introducing comprehensive forest conservation concepts that build on ecosystem persistence and resilience. The fight against illegal logging and mining, and governmental measures against encroachment into indigenous forest land must be part of such a strategy. If we use the lessons from the past twenty-five years of tropical rainforest conservation to show us the way, and take full advantage of REDD+, the potential of slowing and ultimately reversing the course of destruction still exists. Big changes could be in the air.

CLÁUDIO C. MARETTI[1]

The Amazon: There Is Hope!—If We All Do the Right Thing

The Amazon: Some historical glimpses

There are a number of different "Amazonias," depending on whether you use hydrography, ecology, or political definitions to set the boundaries.[2] The most commonly used and least disputed demarcation is the Amazon River Basin, but for many people the most important concept is the ecological Amazon. It is usually defined by the tropical rainforest biogeographic domain, also called the "Amazon biome," and this is the definition used here.

NATURAL FEATURES AND ECOSYSTEM PROCESSES AND SERVICES

One of the most important remaining natural regions in the world, the Amazon is number one in biodiversity. It contains the world's richest diversity of primates, birds, butterflies, and many other groups of plants and animals.[3] Estimations of freshwater fish species continue to grow beyond the current number of about 9,000. To date, at least 40,000 plant species have been identified. But for the largest group of living organisms, the invertebrates, it is impossible to estimate species numbers, so little do we know about them. The region probably includes about 10 percent of the world's known species diversity, but these numbers quickly become outdated because of the diversity of unique and unknown habitats, the inaccessibility of much of the vast Amazon region, and the continuous discovery of new species. Between 1999 and 2013, at least 1,661 new species of plants and vertebrates have been discovered in the Amazon.

The Amazon is also the world's largest river basin. It contains 12–20 percent of the world's fresh water. The Amazon biome drives atmospheric circulation in the tropics by absorbing energy and recycling about half the rainfall. This biosphere–atmosphere "pump" is an important control mechanism for rainfall patterns in other parts of the continent, particularly its central and southern zones.

The natural resources of the Amazon are the basis of many local, national, and even world economies. They are used in food production, as building materials, in tool making, and as sources for producing textiles, handicrafts, pharmaceuticals, biotechnology, dyes, perfumes, resins, gums, oils, and more. They also play a role in sociocultural ceremonies and rituals.

HUMAN OCCUPATION OVER THOUSANDS OF YEARS

Humans have lived in the Amazon region for over 11,000 years. After the arrival of the Europeans, people from Africa were brought in as slaves, and Asians also arrived in the region. Their settlement and exploitation of natural resources had a strong impact on the indigenous peoples through slavery, genocides, and forced settlement. The extraction of rubber catalyzed historically important waves of immigration as the United States and Europe industrialized, the two world wars in the twentieth century created new demands, and the automobile industry grew. Nevertheless, until roughly World War II, the occupation of the Amazon was sparse, with very low population densities and relatively low impacts on nature.

However, since the middle of the last century and up to now, human occupation and use of natural resources in the Amazon have intensified, first in Brazil and later in the other countries that share the Amazon. Geopolitical interests in "national security" and "national integration" have been rising factors. These have entailed road building, new settlements, and military presence. Roads have a particularly strong effect: beyond their direct impact, they open the way for more land occupation and intense use of resources. Exploration and exploitation of minerals, oil, gas, and hydroelectric power indirectly attract more people into the region and promote settlement and the use of natural resources.

Around thirty-three million people are estimated to live in Amazonia today. Human groups inhabit historical cities and industrial zones, living as rubber tappers and riverine settlers, or in African-descended communities. Farmers and ranchers represent an important and diverse group: from descendants of the old settlers to newcomers; from smallholders who came with the agrarian reform to big landowners; and pseudo-ranchers occupying a significant amount of illegally or irregularly acquired land. Some 385 indigenous peoples have been recognized, of which over sixty groups live in voluntary isolation as uncontacted peoples (see plate 16).

In the absence of adequate planning, economic guidance, and government control, an important part of the land occupation by human groups is not regular. It is often associated with violence against the traditional dwellers, prostitution, disease, and slave-like labor conditions. Overexploitation of natural resources, degradation, and deforestation are the consequences. The resistance of indigenous Amazonian peoples and the demand for human rights and nature conservation slowed the occupation process in the 1980s, but in the following decades land occupation increased again, this time on a much more important scale, driven by global markets and the economic growth of the South American countries. The development of hydroelectric power, on hold in Brazil for more than twenty years since the struggle against the dams in the Xingu River basin, has also been taken up again.[4]

The rise of protected areas and indigenous territories

Today, the Amazon is arguably among the best protected of the large and important natural areas on Earth. This important achievement is the result of a combination of factors, including the difficulty of "conquering" the region with the means available in past centuries. It could be called an unintentional result of the Western development model. Another relevant factor has been the deliberate efforts to protect the Amazon because of the significant international awareness of its importance. Protected areas are the best-known mechanism to conserve ecosystems, and if these areas are established correctly, that includes safeguarding the benefits to local people. The conservation of the Amazon has, however, also benefited from the protection of its indigenous peoples and the promotion of sustainable use by local communities.

Indigenous peoples who have inhabited the Amazon for thousands of years in many cases now benefit from social and cultural protection by governments. But the Amazon countries do not all use the same legal systems and policies for indigenous peoples' territories. Governments recognize part of the areas the indigenous peoples still live on as Indian reserves or indigenous peoples' territories. In most cases, indigenous peoples are managing their areas with strong conservation results, either as nature reserves based on a holistic integration of their lifestyles with nature, or by specifically protecting parts of their lands. However, most of their territories are not yet officially recognized as nature conservation areas. Other local communities—including former slaves, fishermen, small farmers, and extractive forest dwellers—also manage their areas for nature conservation.

Although the creation and management of protected areas has not always been respectful of the rights and interests of local communities and indigenous peoples, several governments have increasingly involved local communities' interests in protection strategies. For example, the "extractive reserves" in Brazil are a protected area category created nationally to respond to local communities' needs.[5] This idea influenced the creation and definition of the IUCN protected areas category VI at the international level (for an explanation of IUCN protected areas categories see footnote on page 182).

THE CREATION OF PROTECTED AREAS
SINCE THE 1960S

Until the 1960s, protected areas in the Amazon were mainly established with limited objectives to protect a specific feature, such as a mountain or a set of waterfalls or caves. In the 1970s a significant public opinion movement in favor of conservation and sustainable development emerged, but the growth of land under protection was slow until the late 1980s. The first large protected areas of more than half a million hectares were established in this period—a tendency that continued later on. This size is often considered to be a minimum for consistent biodiversity preservation in the Amazon.[6] The land surface covered by protected areas and the percentage protected of each Amazon country is shown in the figure on page 211.

The increase in the rate of protected area creation has been linked to the evolution of national policies, the evolution of the concept of protected areas, international conventions, and national projects and initiatives. The 1992 Earth Summit in Rio with the Convention on Biological Diversity also had an influence on the motivation to establish protected areas. The Amazon countries instituted legal frameworks with that purpose in mind and defined protected areas categories with complementary management objectives. Subsequently, many governments increased the size and habitat coverage of their protected areas. The Latin American Congresses on Natural Parks and other Protected Areas of 1997 in Santa Marta, Colombia, and of 2007 in Bariloche, Argentina, were particularly instrumental in triggering the creation of new protected areas.

THE AMAZON REGION PROTECTED AREAS
PROGRAM: A BIG LEAP

In the first decade of the new century the expansion of protected areas creation in the Amazon was particularly rapid. In Brazil, the Amazon Region Protected Areas (ARPA) program helped the Brazilian national and state governments create new protected areas and consolidate existing ones. The program was conceived in 1998 after a challenge to the Brazilian government, and its influence became strong after an agreement was signed between the Brazilian government, the World Bank, the Global Environmental Facility, and the WWF at the 2002 Earth Summit in Johannesburg. This program was already ambitious in the planning phase, but it was further enhanced by the inclusion of sustainable use reserves (IUCN category VI) involving local communities. The extended program covered 12 percent of the Brazilian Amazon, completing mosaics of protected areas and adapting them to the needs of local communities.[7]

MOSAICS OF PROTECTED AREAS

Whereas in the 1990s much attention was given to the establishment of large protected areas, in the years after 2000, mosaics of protected areas with various land use definitions were proposed to meet a variety of conservation objectives. Some of these mosaics are particularly significant for the Amazon:

- Around 2004, Peru accomplished a significant mosaic in the Alto Purús national park area. The Purus–Manu conservation corridor, now about ten million hectares in size, includes previously established protected areas. It is important for the protection of indigenous peoples, including some in voluntary isolation.
- The Apuí Mosaic in the Brazilian state of Amazonas was created in 2005. Together with the creation of the Juruena and Campos Amazônicos National Parks in 2006, and complementing a series of other protected areas, this complex forms part of the Southern Amazon Mosaic—an area of more than seven million hectares (see plate 17).
- Brazil made significant progress with the Terra do Meio Mosaic in Pará in 2005–06, which covers more than eleven million hectares. The Terra do Meio Mosaic is adjacent to indigenous territories, complementing the conservation of an area larger than twenty-five million hectares, a large part of which falls in the Xingu River basin.

- A landmark achievement was the creation of the Calha Norte mosaic of Pará in 2006, complementing large existing protected areas, in total an area of almost thirty million hectares. This mosaic connects to protected areas of the Guiana Shield countries and is likely to constitute the largest protected pristine tropical rainforest area of the world.
- In 2008–09 the federal government and the state government of Amazonas put 8.4 million hectares under protection, joining it in a mosaic with other preexisting protected areas to mitigate impacts from the BR-319 highway that connects Porto Velho with Manaus.[8]

As part of a larger strategy the creation of protected areas was increasingly planned to contain deforestation. Several of the mosaics followed a vision of "green barriers" against the advance of the deforestation front (see plate 8).

CURRENT PROTECTED AREAS COVERAGE

By 2010 the extent of the Amazon under protection had grown very significantly. About 166 million hectares, in 401 protected areas, representing 25 percent of the Amazon was now protected. This corresponds to an area almost as large as Germany, France, Spain, and Italy together—clearly a great global achievement. There are considerable differences from country to country in the percentage of rainforest protected: while Guyana only has 2.2 percent of its tropical rainforest included in its protected areas system, French Guiana has almost half its territory protected. Most of the Amazon countries, though, have protected far more of their tropical rainforests than the global average. Several countries (Brazil, Bolivia, Venezuela, Ecuador, and Peru) have between 20 and 30 percent of their share of the Amazon Basin in protected areas[9] (see figure and table on pages 211 and 212).

Protected areas in the Amazon

Changes in the amount of protected forest area in the Amazon between 1960 and 2010 (*above*) and percentage of each country's share of the Amazon in protected areas (*below*). Indigenous territories are not included. Adapted from Riveros et al. (2014)[10]

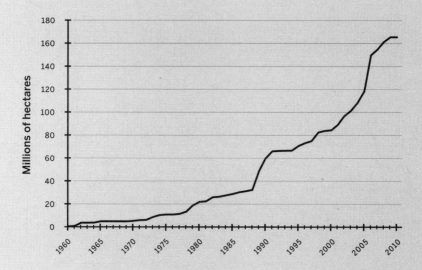

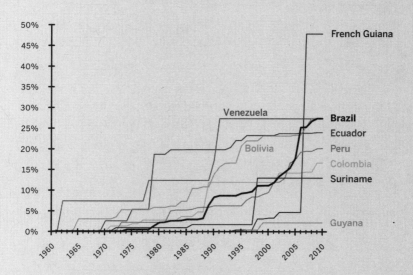

Total area and number of protected areas in the Amazon established from 1960 to 2010

Source: Maretti et al. (2012), including adaptations from Riveros et al. (2014).[11]

	TOTAL AREA PROTECTED (HA)		NUMBER OF PROTECTED AREAS		% OF AMAZON SHARE PROTECTED
Bolivia	12,183,719	7.3%	39	9.7%	27.4%
Brazil	110,413,575	66.3%	243	60.6%	27.4%
Colombia	8,089,354	4.9%	18	4.5%	16.6%
Ecuador	2,876,324	1.7%	14	3.5%	24.0%
Fr. Guiana	3,904,467	2.3%	11	2.7%	47.8%
Guyana	455,881	0.3%	4	1.0%	2.2%
Peru	15,885,858	9.5%	35	8.7%	20.1%
Suriname	1,946,727	1.2%	12	3.0%	13.8%
Venezuela	10,750,086	6.5%	25	6.2%	27.3%
TOTAL	166,505,994	100.0%	401	100.0%	24.9%

It is interesting to note the preferences of the different countries for IUCN categories of protected areas. Brazil has the majority in category VI—"sustainable use of natural resources"—both in number of areas and in total area. In fact there are several categories in the Brazilian national legislation corresponding to this category.[12] In the past decades, there have been two important tendencies in how protected areas are categorized and managed:

1) More attention has been paid to building constellations of protected areas, both in geographical and management terms, which may take the form of corridors, mosaics, or "green barriers."

2) Governments and the conservation movement have shown a more inclusive approach to integrating the rights and interests of local communities and indigenous peoples. Indigenous territories have also increasingly been recognized as being of great value to nature conservation.[13]

INDIGENOUS TERRITORIES

There are important differences between countries in how they recognize the legal rights of indigenous peoples, and their governments' procedural capacity for communicating with them. There are so many indigenous groups and territories that it is virtually impossible to obtain a complete record. According to my information, in 2010 there were more than 3,000 indigenous territories in the Amazon, although not all of them were officially recognized. The total area of these territories covered almost 208 million hectares. Peru had the largest number of areas (many of which are not fully defined yet), whereas Brazil had the largest area included in indigenous territories. Venezuela had the largest percentage of its share of the Amazon declared as indigenous territories (see table below).[14] The size of indigenous territories varies significantly: Venezuela has some of the largest ones (more than 0.5 million hectares) followed by Brazil, and Peru is at the other extreme with an average size of little more than 10,000 hectares.

Amazon indigenous territories by country

Adapted from RAISG (2010).[15]

	NO. OF ITS	TOTAL AREA (HA)	AVERAGE SIZE (HA)	% OF COUNTRY'S AMAZON SHARE
Bolivia	393	14,239,720	36,233	32.1%
Brazil	312	100,889,351	323,363	25.1%
Colombia	185	25,326,969	136,903	51.9%
Ecuador ·	357	7,664,613	21,470	63.9%
Fr. Guiana	22	715,105	32,505	8.8%
Guyana	116	3,167,084	27,302	15.0%
Peru	1,581	20,622,634	13,044	26.2%
Suriname	23	4,918,469	213,846	34.8%
Venezuela	54	30,380,355	562,599	77.2%
TOTAL	**3,043**	**207,924,300**	**68,329**	**31.1%**

Even if the primary purpose of indigenous territories is to recognize indigenous peoples' rights to land and natural resources, to respect cultural diversity, and to allow them to develop in their preferred way, these areas also contribute significantly to nature conservation goals. Most indigenous peoples treat their land's resources in an integrated way with their cultural, social, and economic activities. Having the option to live as much in harmony with nature as desired, they often set aside parts of their lands and manage large areas sustainably, with very low intensities and impacts.

Nature conservation includes ecological representation

Even though protection in the Amazon is much better developed than the world average (see plate 14), it still falls short of many conservation objectives. We still need to pay more attention to ecological representation to ensure a full array of ecosystem services for local communities and indigenous peoples, freshwater conservation, biodiversity, and climate change mitigation and adaptation.

The Convention on Biological Diversity Aichi biodiversity targets (see box 9.1) require at least 17 percent of terrestrial and inland water areas of particular importance to be conserved until 2020. However, the Amazon counts thirty-five terrestrial ecoregions and not all of them are sufficiently represented in protected areas. The Brazilian biodiversity conservation target of 2010 aims to protect 30 percent of its Amazon forests, and the WWF considers such a percentage for all Amazonian ecoregions to be the minimum for preserving the biodiversity of the richest biome on Earth.[16]

Creating new protected areas might be feasible for certain ecoregions but much more difficult for others. To maintain the ecosystem services from the Amazon, particularly climate stability, 60–70 percent of the region needs to be kept under an intact vegetation and rivers structure and with ecological processes functioning. Land conversion should not exceed 20 percent of the total biome area. Given that indigenous territories cover a significant part of the Amazon, there is a need to develop innovative alliances for the benefit of the Amazon ecosystem and its original inhabitants so that conservation goals can be met while the indigenous people maintain control of their land.

Protected areas are defined and created to be permanent. Obviously some flexibility is necessary to adapt to new requirements and correct mistakes. In the Amazon, this can still be done with a minimum of negative impacts. If the protected areas are seen as part of a system that provides services to local and indigenous communities, and makes a contribution to the sustainable development of a region, such adaptations can be meaningful. But any proposals to downsize or delist protected areas should undergo technical analysis and political negotiation at least at the same level that was necessary to create them. Compensation must be considered to maintain biodiversity values and ecosystem services. Unfortunately this is not what has been happening in the Amazon over the past few years. The Amazon is suffering from increasing pressure from several fronts, and this includes threats to protected areas without compensation. Political and physical threats are also escalating against the indigenous territories.[17]

DEFORESTATION, PROTECTED AREAS, AND INDIGENOUS TERRITORIES

As tropical deforestation accounts for a large part of global greenhouse gas emissions, the benefit of protected areas for the reduction of deforestation is of particular interest (see plate 8). Although there is always criticism about the effectiveness of protected areas, studies have shown that deforestation rates inside protected areas and indigenous territories are less than a thirtieth of what they are outside.[18]

Even though indigenous territories are not defined as nature conservation areas, much less set up to specifically address deforestation, it has been demonstrated that they nevertheless resist deforestation.

The presence of indigenous peoples, as well as governmental authorities, is an important deterrent to deforestation by illegal loggers,
miners, and other outsiders. Attacks on indigenous territories might
lead to fines and prosecution, and some indigenous groups have developed specific tactics to protect the limits of their lands. The purpose
of protected areas established for biodiversity conservation is not
much different from the protection of indigenous territories, except
that some of them are uninhabited. Land titling and land use definition
for indigenous territories, including the necessary enforcement, may
therefore be effective in keeping away land-grabbing interests and a
tool to lower deforestation pressure.[19]

Which Amazon in 2050?

Of all the regions on Earth, the Amazon is one of the best placed to
become a true sustainable development model by 2050. Decision makers from the whole world, from the municipal leaders of countryside
Suriname to the traders in China and Japan; from the indigenous peoples in the Bolivian Amazon to the energy consumers in São Paulo and
the United States; from those responsible for protected areas in the
Venezuelan Amazon to the private and public banks in Europe: they all
have a responsibility to avoid the damage that the destruction of the
Amazon brings to its people and the whole world. Here is a set of strategies to follow:

• Recognize the rights of the Amazon's indigenous peoples to their land
and its natural resources (particularly those that do not yet have legal
rights). This should include prior informed consent for activities that
affect them or their land.

• Recognize the rights of other local communities, particularly the more
traditional ones and those that live in close interdependence with the
ecosystem, and promote their economic integration while safeguarding
their way of life.

- Keep up the pace of protected areas creation with the goal of reaching 30 percent of a true ecological representation. Improve the management quality and effectiveness of the existing ones, and manage them as systems of ecological networks.
- Integrate protected areas in economic policy, and account for the services they provide to societies. Assure a mechanism for the financial sustainability of the protected areas systems. Avoid downsizing and delisting protected areas and indigenous territories without scientifically sound and socially accepted processes of compensation.
- Adopt national policies and programs to control and avoid deforestation aiming at zero net deforestation and degradation by 2020, and limit the ecosystem conversion in the whole Amazon to a maximum of 20 percent.
- Support science for environmental monitoring, including monitoring of ecosystem conversion by deforestation and forest and freshwater ecosystem fragmentation.
- Establish better regulation and voluntary standards for public and private economic activities and investments. This should include biodiversity and greenhouse gas offset procedures that consider the full range of the services provided by the affected ecosystems.
- Use science and technology to develop techniques to replace slash-and-burn agriculture, avoid interrupting rivers for hydroelectric power production, and prevent disruptive road access to remote areas.
- Promote clarity in land tenure systems and access to natural resources. Establish rules for natural resource rights on collectively owned land, privately owned land, and possible combinations.
- Promote cooperation and integrate policies and programs between the nine Amazon countries with the common goal of preserving the integrity of the Amazon rainforest.
- Ensure democratic and participatory governance in the entire Amazon region.

- Create attention and care for the Amazon as a globally important region by governments, companies, and societies, while respecting national sovereignty and the rights of the local communities and indigenous peoples.

  Taking care of the Amazon is a matter of global importance. Global changes are increasingly affecting this region, and the consequences of its degradation will affect us all. For the Amazon countries, their local and indigenous communities, and the whole world, caring for the Amazon means also taking care of themselves—ourselves.

CLÁUDIO MARETTI *is the leader of the Living Amazon Initiative, a global initiative of the* WWF *Network.*

8

CLIMATE CHANGE IN THE
TWENTY-FIRST CENTURY
The End of Rainforests?

ROPICAL DEFORESTATION AND forest degradation are currently estimated to contribute between 10 and 12 percent of global greenhouse gas emissions.[1] This proportion is less than previously estimated because of the reduction of deforestation in Brazil and because the global emissions from fossil fuel combustion have been rising. However, the remaining forest is increasingly affected by climate change, and various climate models suggest that they could turn from a carbon sink into a carbon source in the twenty-first century. If this happens, we will not only lose a very large part of the Earth's biodiversity, but greenhouse gas emissions will accelerate with no end in sight. An important scientific undertaking is to determine the conditions under which forests could turn into net carbon emitters by studying the response of tropical forests to climate variations of past millennia.

Why are the climatic variations of thousands or millions of years ago relevant to the future of today's rainforests? Using climate models to project the future response of species and ecosystems is a relatively uncertain business. Climate models are notoriously uncertain, and this is compounded by our lack of knowledge about how climate change will affect plants. We have a better scientific foundation for our knowledge about the response of the tropical forests to paleoclimatic events during the Pleistocene and Holocene epochs (2.5 million years ago to 11,700 years ago and 11,700 years ago to the present respectively). Ice cores provide a precise record of climate variations reaching back as far as 800,000 years, and analysis of pollen and other fossil records provides information about how the vegetation responded. As some of the conditions during this period resemble the probable climatic conditions under projected atmospheric carbon dioxide concentrations of the coming decades, we can make predictions on a less hypothetical base.

The legacy of paleoclimates

FOSSIL RECORDS SHOW that the Earth's climate has never been stable over longer geological periods. Between 65 million years ago and 1.8 million years ago, tropical rainforest areas contracted during the drier periods and were replaced by grasslands or drier forests, and the rainforest plant life was impoverished.

THE AFRICAN RAINFOREST REFUGIA

THESE DRY PERIODS seem to have been particularly severe in Africa, as we know from the paleoenvironmental research of French palynologists (scientists who study spores and other microparticles in fossil sediments) Raymonde Bonnefille and Jean Maley who carried out extensive research of pollen sedimentation across Africa.[2] Because African climates are drier and their rainforests are closer to the physiological limits of moist forest vegetation, they are more vulnerable to dry periods. The somewhat lower plant diversity in African rainforests than in Asian and South American rainforests is a consequence of these dry periods. Particularly the laurel (Lauraceae), palm (Palmae), and arum lily (Araceae) families are considerably less diverse in Africa, although African rainforests generally have a higher biomass per area and trees with more impressive stem diameters and buttresses. Many African rainforest areas today still receive less rainfall than Southeast Asian rainforests or some parts of the Amazon.

The sensitivity of African rainforests to dry periods makes it especially interesting to study their paleoclimatic response. There is a lot of evidence that rainforests in Africa retracted to relatively small areas (refugia) during the more recent glacial periods of the Pleistocene epoch when the climate in equatorial regions was cool and dry, which did not favor tropical rainforest growth.

With the increasing concern about the impact of climate change on forests, the refugia theory, previously used to explain the uneven distribution of rainforest species and their evolution, has attracted new attention. A number of recent scientific studies use our knowledge of paleoclimatic conditions and their impact on African forests to project the effects of warming climates in this century.[3]

Popular articles commenting on the dire fate of tropical rainforests sometimes refer to the "millions of years" these forests have existed, but for many rainforest areas nothing could be further from the truth. In fact some of them may not have existed during the last glacial period, which reached its maximum around 20,000 years ago and ended only about 12,000 years ago. Geomorphological evidence suggests that Africa's climate may have gone through severe cool and dry periods in the mid-Pleistocene, when the Sahara and Kalahari deserts extended to within one or two degrees of the equator and destroyed almost all of the African rainforests.[4] Kalahari sands under the thin topsoil layer of the southern part of the Congo Basin forests are evidence that sand dunes blew across land that is covered by rainforests today.[5]

Although the size of the former African rainforest refugia is uncertain, the current distribution of plant and animal species is testimony that they once existed. Rainforests expanded again after the end of the last glacial period, but some tree and animals species lagged behind in re-expanding their range.[6] The refugia during the last ice age seem to have been located in the areas of highest rainfall on the West African coast, in current Sierra Leone and Liberia, and in western Cameroon, as well as in the eastern Democratic Republic of the Congo and on mountain massifs. These regions still have higher rainfall than the rest of the current African rainforest area.

The former African forest refugia are recognizable as "centers of endemism" that have a higher species diversity than surrounding areas. Out of all the forest mammals in Africa, 63 percent are found within these centers of endemism. Today it is generally believed that the isolation of these African forest refugia during the Pleistocene did not play an important role in the speciation of the forest flora—for example the evolution of distinct tree species. The rainforest flora originated much earlier, perhaps 70 million years ago, since when it has become modified and diversified.[7]

The African forest refugia of the Pleistocene, on the other hand, may well have promoted the speciation of certain animal groups, for example the several species of black-and-white colobus monkeys or certain butterfly species whose distributions are close to these former forest refugia.[8] Other scientists have questioned whether current species distribution patterns can reflect former forest distribution after thousands of years of complex paleoclimatic variations. The fact that many plant and animal species share a coevolutionary history and therefore remain dependent on each other could help explain the very slow expansion of some species' ranges, and the patchiness of their current occurrence. Coevolutionary, symbiotic relationships between plants and animals are a more credible explanation for the current structure and biology of African forests than the often repeated conjectures about early human influence.[9]

THE HOLOCENE IN AFRICA: WARM AND HUMID, THEN WARM AND DRY

WITH THE END of the last glacial period and the beginning of the Holocene period, about 11,700 years ago, the African forests began to expand again. Paleontological records of the Holocene may also give a hint of possible rainforest responses to future climates. After the end of the last glacial period, a warm, humid period (the African Humid Period) began in western and eastern Africa and the Sahara and lasted from about 11,000 to 4,000 years before present. It led to an expansion of rainforests farther north. Individual forest trees penetrated the Sahara, which became partly covered by savanna, grass steppes, and a system of interconnected lakes, rivers, and inland deltas. Savanna animals roamed across the Sahara, and even fish species and the open water–dependent hippopotamus dispersed across the Sahara during this period.[10] A warm, dry period followed the African Humid Period from 3,200 to 1,000 before present, which led to a drastic shrinkage of the forest area, although in a nonlinear pattern that left forest fragments and gallery forests (forests that grow along the banks of watercourses) behind. Slow climate trends in certain locations seem to have triggered abrupt vegetation changes, and rapid climate trends caused slow vegetation response in others, depending on the local soil and geomorphology.[11]

PALEOCLIMATIC LESSONS FROM AFRICA: THREAT OF EXTENDED DRY SEASONS

TAKING THE CONSEQUENCES of the climatic variations during the glacial periods of the Pleistocene and the warming in the Holocene into account, what effects is the forecasted warming in Africa expected to have? The forest shrinkage during the relatively recent warm and dry period that started about 3,200 years ago is of particular relevance to this question. The International Panel on Climate Change's General Circulation Models indicate that Africa will experience between 3°C and 4°C of warming by the end of this century. The last time such rapid warming happened was toward the end of the last glacial period, between 14,700 and 11,500 years ago, when two abrupt warmings took place within only a few years, as the North Greenland ice core research has revealed.[12]

It is not clear how sensitive plants and animals will be to the projected temperature increase. Nor is it possible to project the effects of higher atmospheric carbon dioxide concentrations, which has a fertilizing effect on plants and which will be considerably higher than during the glacial periods of the Pleistocene. The prolonged and intensified dry seasons most climate models predict are likely to have a much bigger effect on the tropical rainforest.

Take the western Congo Basin. Drying climates in West and Central Africa are associated with warming sea surface temperatures in the tropical North Atlantic and a northward shift of the intertropical convergence zone, which is responsible for the West African monsoon.[13] As the African rainforests are generally drier and closer to the physiological limits of rainforest vegetation, and also have a proven paleoclimatic track record of converting to savannas, extended dry seasons and more severe droughts are likely to put large rainforest areas in danger. In fact, the 4°C warming scenario predicts such a big rainforest retreat in the western Congo Basin that the forest area would almost be split into a western and an eastern block,[14] reminding us of the forest refugia of the last glacial period.

RESPONSE TO PALEOCLIMATES IN OTHER REGIONS

IN SOUTHEAST ASIA, paleontological records have also provided evidence of tropical rainforest refugia. The dynamics of sea-level rise and fall during the glacial periods of the Pleistocene, though, were complex in this part of the world, with frequently changing shorelines. Solid evidence exists that during the last glacial maximum, about 20,000 years ago, when sea levels were at least 120 meters lower, a vast continental land area ("Sundaland") connected the Malaysian Peninsula with Sumatra, Java, and western Borneo. A savanna corridor of undetermined extent separated the rainforest refugia in the higher elevations of today's Sumatra, Java, and Borneo.[15] This savanna corridor is still a hotly debated issue, but it may also have served as a dispersal route of *Homo sapiens* to Java and beyond, although there may have been earlier dispersals of man during the glacial periods of the Pleistocene, some 40,000–60,000 years ago.[16] Generally speaking, however, the Southeast Asian rainforests have been remarkably resilient to the climatic variations of the Pleistocene and more recent times, not least because of the highest sea surface temperatures on Earth and high rainfall in this region as a result.[17]

The theory that tropical rainforests withdrew into refugia surrounded by grassland areas during glacial periods has not been confirmed by paleontological records for the Amazon. Lowland forests may have persisted in the Amazon Basin, but as a drier, more seasonal forest type. The warm, dry climatic conditions during the earlier to middle Holocene (about 8,000–4,000 years ago) affected the margins of the rainforest area in eastern and southern Amazonia, where forests were replaced by open savanna. At least in the most recent scientific literature, no evidence can be found for a widespread expansion of savanna across the Amazon Basin. However, deposits of pollen from *Cecropia sp.*, a fast-growing secondary forest tree associated with forest disturbance, provide some clues. High peaks of *Cecropia* pollen in lake sediments point to more open vegetation caused by drought and fires, including fires caused by Paleo-Indians.[18]

Dangerous rainforest tipping points ahead?

WHILE THE PLEISTOCENE climate variations were probably caused by a combination of ocean circulation and solar energy fluctuations, today's rapid climate change is overwhelmingly anthropogenic. The current rate of greenhouse gas emissions, the lethargy of governments, and the impotence of the intergovernmental system make it very unlikely that average global warming will be kept below 2°C. Under a midrange emissions scenario, atmospheric carbon dioxide concentrations will probably rise by the end of this century to levels never experienced in the last fifty million years, and temperatures could increase 3°C–4°C, or even higher under a widespread forest destruction scenario.[19]

We therefore have to expect tipping point effects—abrupt changes in the condition of many ecosystems. After the polar regions, which are exposed to a more rapid than average warming rate, and coral reefs, which are particularly sensitive to sea water warming, tropical forests are the third group of ecosystems that may fall victim to tipping point effects. A recent report of the US National Research Council dealt with the question of whether such sudden surprises could be anticipated and mitigated.[20]

The paleontological records show that climate change manifests as changing temperatures and precipitation patterns, as well as more marked seasonality with longer drought periods. There has been concern that these changes could cause a forest die-back in the coming decades, further reducing the area of tropical forests and thus their capacity to absorb atmospheric carbon dioxide. Forest die-backs, defined as tree mortality noticeably above the usual mortality level, have in recent times been documented mainly in arid and semi-arid regions where trees are at the limit of their physiological tolerance.[21] However, these symptoms could foreshadow what may happen to tropical rainforests to a much larger degree over the next decades. Widespread climate-induced changes in tropical forest ecosystems could change them from net carbon sinks to massive sources of atmospheric carbon within this century, which would speed up further global warming. It has even been suspected that such a feedback loop could turn the terrestrial biosphere into a carbon source after about 2050.[22]

The remaining tropical rainforests are already being altered through climatic factors that affect their structure, dynamics, and productivity. Simon L. Lewis, a plant ecologist at the University of Leeds, identified plausible feedback scenarios that could turn tropical forest into net carbon dioxide producers.[23] Although plants initially grow faster because the higher carbon dioxide concentration increases the photosynthetic rate, biomass production is soon limited by the scarcity of soil nutrients and water. At the same time, higher temperatures increase plant respiration: the inverse process of photosynthesis, respiration consumes energy and produces carbon dioxide. As temperatures continue to rise, increasing respiration costs will exceed photosynthetic rates, making the remaining tropical forest a net source of carbon dioxide.

DRY SEASONS VERSUS TOTAL RAINFALL

TROPICAL RAINFORESTS ARE characterized by high annual precipitation patterns, generally above 1,500 millimeters, relatively evenly distributed throughout the year. More important than the total rainfall in a year is the strength of the dry season. The duration of the dry season and water stress have the strongest influence on rainforest distribution. The "maximum climatological water deficit"—the accumulated deficit built up when water received from precipitation does not replace water lost through evapotranspiration during a year—has the strongest explanatory power for the presence or absence of rainforests. Total annual precipitation is the second most important factor after these water stress factors, followed by temperature and some other factors.[24] Evaluating the impact of climate change on rainforests, therefore, means focusing on the length of dry seasons and water stress more than temperature.

THREATS FROM INCREASING DROUGHT FREQUENCY

DROUGHT, WITH THE resulting forest fires, remains the most likely cause of forest collapse and severe tipping point risks. Droughts in tropical rainforests have become more frequent and severe in the past two decades. However, until recently little was known about the complex relationship between drought periods, net primary productivity—that is, tree growth—drought-induced tree mortality, and forest fires. It is in this context that the future of the tropical rainforests, under business-as-usual climate change scenarios, will be determined. The effect of drought on rainforest ecosystems has now become a major research focus, particularly as it concerns the Amazon Basin.

But before considering the consequences of drought in a time of changing climate, we should understand what causes droughts. When the twentieth century's strongest El Niño–Southern Oscillation (ENSO) event occurred in 1997 and 1998, it was considered to be an unusual phenomenon, although El Niño had been known for centuries and the phenomenon has been scientifically described since the 1930s. The 1997–98 event caused severe drought in Amazonia, Southeast Asia, and Mexico and had massive effects on the net primary productivity of forests, thus their capacity of carbon storage, as well as forest fires. Only three years later, in 2001, another severe drought occurred following an ENSO event. About one-third of the Amazon forests became prone to fires, and stored carbon at a significantly lower rate.[25] In 2005 another drought occurred, which was then considered to be a once-in-a-hundred-years event, followed by yet another in 2010!

TWO FACTORS INFLUENCE the frequency and severity of drought:

1) The increasing sea surface temperatures in the Pacific Ocean enhance the periodic El Niño effect and inhibit convection in northern and eastern Amazonia.

2) High Atlantic sea surface temperatures with a northwestern shift of the intertropical convergence zone trigger more frequent droughts. Warmer north Atlantic sea surface temperatures strongly influence the dry season length and intensity in southern and eastern Amazonia.[26]

DROUGHTS TURNING AMAZONIA INTO A NET CARBON DIOXIDE SOURCE

THE AMAZONIAN RAINFORESTS have been considered to be particularly vulnerable to tipping point effects in response to climate change combined with forest conversion. Some Amazonian rainforest areas receive less than 1,500 millimeters of precipitation, which makes them more likely to have long dry seasons and consequently decreases their carbon storage capacity. In normal years, rainfall is roughly equal to evapotranspiration. But during ENSO years, rainfall drops below evapotranspiration levels and gradually depletes soil moisture, sometimes to a depth of more than ten meters.

As global warming effects become stronger, ENSO events become more frequent, and rainfall is further inhibited by forest loss and fragmentation, droughts are likely to become more common and more severe.[27] The 2005 and 2010 droughts in the Amazon Basin (see plate 15) caused a loss of carbon storage of 1.6 to 2.2 gigatonnes (billion metric tons) of carbon, calculated as the loss in biomass accumulation caused by stress on the trees, plus the carbon released by decomposing dead trees. These effects turned the Amazon forest into a carbon source during these drought periods! Two such drought events can offset the carbon storage (approximately 0.4 gigatonnes of carbon per year) of a whole decade.[28]

FLAMMABILITY ON THE INCREASE

GENERALLY DRIER CLIMATES and longer dry seasons make tropical forests susceptible to fires caused by lightning and human activity. During the ENSO event in 1997–98 it was estimated that more than 20 million hectares of tropical forest, about the area of Great Britain, was burned.[29] Although such fires are normally triggered by slash-and-burn and other land clearance activities, under drought conditions they risk destroying much larger areas as they spread as ground fires into surrounding forest areas. A particularly disastrous fire during the 1997–98 ENSO event happened in Kalimantan, Indonesia, when large areas of the peat swamp forest burned and the smoldering peat released large quantities of carbon into the atmosphere. Over the whole of Indonesia, forest fires contributed between 0.81 and 2.57 gigatonnes of carbon emissions, between 13 and 40 percent of the global emissions from fossil fuels in 1997. It caused the largest annual increase in carbon dioxide concentration in the atmosphere since measurements began.[30]

What seemed unlikely until a few decades ago, that rainforests could be exposed to forest fires, has now become a major threat. Intact humid evergreen forests are certainly far less prone to fires than drier forest types, and when they are continuous and undisturbed, they are remarkably resistant to fires even in drought conditions. A closed forest canopy traps the moisture that the trees and other plants lose through transpiration and creates a humid environment that prevents fires from spreading.[31]

However, once a rainforest area has been opened up, degraded, or fragmented, the edge effects come into play. Humidity levels decrease with open canopies, and the dry leaf litter becomes susceptible to fires spreading from surrounding cleared areas and fires caused by slash-and-burn activities. Although such ground fires are "cold" fires, which means they burn at a low intensity, they can nevertheless cause important tree mortality, because rainforest trees have thin bark and are generally not fire resistant. Ground fires also destroy the seed stock and tree seedlings on the ground, reducing the forest's ability to regenerate.

Selective logging makes forests more vulnerable to ground fires until secondary vegetation has overgrown the forest gaps created by felled trees, logging roads, and loading places. Low-impact logging techniques can significantly reduce these risks.[32] Even without fires, fragmented forests have higher tree mortality near the fringes through desiccation penetrating several hundred meters into the fragments.[33]

Climate model uncertainties

SOME ASPECTS OF how climate change will affect the future of rainforests are still uncertain. A temperature rise alone seems unlikely to reduce the extent of rainforests in the twenty-first century. New areas may even become suitable for tropical rainforests.[34] Large-scale climate models are known to have important areas of uncertainty, particularly regarding changing rainfall patterns.

One question concerns the possibility of a large-scale Amazonian forest die-back. This appears to be a particular risk in southern Amazonia, exacerbated by the forest fragmentation found there. The HadCM3 model developed by the United Kingdom's Met Office Hadley Centre, which predicts 4°C of global warming, implies that more than half the Amazonian forest will retreat by the end of the twenty-first century,[35] whereas other models predict a loss of only 10 percent. Central America and the Caribbean also show a high probability of rainforest retreat due to climate change. The HadCM3 model was based on a business-as-usual emissions scenario[36] and predicted a strong feedback loop between forest loss, higher carbon dioxide emissions, and lower sequestration capacity. It is now considered an extreme scenario. Divergences in modeling arise out of uncertainty about how Pacific and Atlantic ocean surface temperatures will react to warming trends. After the Hadley Centre revised its calculations with a weaker feedback from higher carbon dioxide emissions, the new model, HadGEM2-ES, showed much less climate-induced forest loss by the end of the century.

Nevertheless, the earlier projections of extensive Amazonian forest die-back remain a possible outcome.[37] The uncertainties of climate models should not distract us from the current trends of increasing drought frequency and severity, which show a much faster pace than climate models would predict. If droughts like those of 2005 and 2010 become the norm in the coming decades, some models predict a large-scale replacement of the Amazon forests by open forest types and savanna-like vegetation by the end of the twenty-first century.[38] Since 1979, the dry season over southern Amazonia has become longer, which means a longer annual fire vulnerability period.[39] With the combined effects of higher emissions, drying, forest fragmentation, and fires in tropical rainforests, there is a serious risk of a tipping point, a self-reinforcing effect of forest die-back, with catastrophic consequences for life on Earth, including its human inhabitants.

Whether we are urban or rural citizens, as estranged from nature as we may be, we all depend on the Earth's ecosystems for food, water, and many other ecosystem services, and one of the ecosystems we depend on is the world's tropical rainforest. Continental rainfall patterns are largely determined by the evapotranspiration of the large rainforest blocks: air that has passed over large forest areas produces at least twice as much rainfall in the region as air that has passed over little vegetation.[40] The Committee on Abrupt Climate Change of the US National Research Council concludes in its most recent assessment that "credible possibilities of thresholds, hysteresis, indirect effects, and interactions amplifying deforestation make abrupt (fifty-year) change plausible in this globally important system."[41]

Droughts and fires exacerbate risks for rainforests

RESEARCH INTO THE effect of meteorological, hydrological, and ecosystems changes on tropical rainforests increasingly shows that anthropogenic greenhouse gas emissions have a major impact. The effects are not likely to be directly caused by higher than average temperatures, at least not in the next decades. Instead, higher temperatures translate into major meteorological changes through warming sea surface temperatures. These in turn change seasonal rainfall patterns in rainforest areas, causing longer and more intense droughts, depleting soil moisture, and resulting in fires.

The complex interrelationships between global warming, rainfall patterns, drought, fire hazards, and land use change in tropical forests require more detailed assessments and research, but it can already be said that a gradual die-back of rainforest areas with less than 1,500 millimeters of rainfall in Africa and the Amazon combined with increasing fire hazards is a realistic scenario for the rest of this century. Water—predominantly the even, seasonal distribution of rainfall—defines a rainforest with all the life in it. When this pattern gets disrupted recurrently over decades, the future of the rainforest is in grave jeopardy.

Climate change further amplifies the existing threats to rainforests from deforestation and degradation. Fragmented forests are particularly vulnerable to longer and more frequent drought periods, and global warming models for the twenty-first century show an increased risk of drought in the next thirty to ninety years resulting from decreased precipitation, increased evaporation, or both, especially over most of the Americas and southern and Central Africa.[42] Forest fires across the humid tropics will therefore increase in proportion with the spread of vulnerable forest area.[43]

We can only guess what this means for biodiversity on the global scale, but such research programs as the Biological Dynamics of

Forest Fragments Project (see box 6.1) or the Living Planet Index (see figure 6.1) provide an alarming preview. We also know that with increasing droughts in rainforest areas, the carbon storage capacity of forests is greatly reduced, and we have already seen years with El Niño effects when Amazonia was a net carbon source. It is not exaggerating to say here that decision makers are tampering with dangerous feedback loops when they fail to take serious action to mitigate greenhouse gas emissions. We are not far from the tipping point of ecosystem collapse in some of the most important rainforest areas of the world!

CHRIS ELLIOTT

REDD+ Is a Little More Complex than "Not Cutting Down a Tree"

The origins of REDD+—from Bali to Cancún

When REDD (Reducing Emissions from Deforestation and Forest Degradation in Developing Countries) was first discussed in the mid-2000s, the Intergovernmental Panel on Climate Change estimated that tropical deforestation and degradation had contributed about 25 percent of global greenhouse gas emissions in the 1990s.[1] Since then the proportion has decreased, mainly because of an increase in fossil fuel consumption, but partly because of a major reduction of deforestation in Brazil. The current figure is estimated to be 15 percent, which is 1.4 billion metric tons of carbon per year,[2] still a very significant amount.

REDD was proposed in its current form by the Brazilian researcher Santilli and colleagues in 2005.[3] The term used was "compensated reduction" whereby countries that committed to reducing their national deforestation rates below a previously determined historical level would receive international compensation should these reductions be sustained. Compensation would come from the sale of emissions reduction certificates to governments or private investors. The concept of REDD was taken up by environmental groups and some tropical countries, the latter informally organized into a Coalition for Rainforest Nations, and in 2007 at the United Nations Climate Change Conference in Bali, Indonesia, the Bali Road Map was adopted as a two-year plan to finalize a binding international agreement in December 2009 in Copenhagen. The road map[4] included a reference to the concept, by then called REDD+ (the plus sign signified the inclusion of reforestation and forest restoration) as follows:

Policy approaches and positive incentives on issues relating to reducing emissions from deforestation and forest degradation in developing countries; and the role of conservation, sustainable management of forests and enhancement of forest carbon stocks in developing countries.

During this period there was considerable enthusiasm about REDD+ in climate change circles as it was seen, in some cases rather naively, as a quick and cost-effective way of reducing greenhouse gas emissions. The influential report by Nicholas Stern for the United Kingdom government estimated the cost of implementing REDD at USD 5 billion a year initially.[5] Another estimate put the costs of reducing deforestation by 50 percent by 2030 at USD 17–33 billion per year, assuming forests were included in a global carbon trading system.[6] Both reports have been criticized for using an "opportunity cost" approach, which underestimates implementation costs, rather than looking at the real costs of reducing deforestation.

Over the next three years negotiators made slow but significant progress on REDD+ at the meetings of the United Nations Framework Convention on Climate Change (UNFCCC), and although the Copenhagen meeting in 2009 was a disappointment in many ways, a decision was finally reached in December 2010 in Cancún. It should be noted that one positive outcome of Copenhagen was the commitment of USD 4 billion of "fast start" funding to support REDD+. Even though a portion of these funds was "old wine in new bottles," that is to say funds that had been committed already and were now being labeled REDD+, some of the funding was new. The new funding included a significant commitment from Norway, which was emerging as a major supporter of REDD+. The major outcome of Cancún was guidance for tropical countries on preparations for REDD+. This led to significant media coverage, including a special issue of the *Economist*.[7] The agreement recognized that a phased approach needs to be taken before payments for emissions reductions can be made. The importance of safeguards to ensure that REDD+ payments do not have a negative effect on local communities or indigenous peoples was also recognized.[8] A lot more work was required after Cancún but the key building blocks for REDD+ were eventually completed at the UNFCCC meeting in Warsaw in 2013.

The building blocks

Part of the success of REDD+ is that it is a broad enough concept that it can mean different things to different people. This has also allowed it to evolve in response to external events.[9] Its initial focus was on carbon offsets, projects, and private-sector funding, but when this did not materialize the discussions shifted to national and subnational REDD+. Although the initial objective of REDD+ was to reduce carbon emissions, over time the focus broadened to include maintaining other forest functions and securing land rights for local communities.

Despite the evolving objective of REDD+, some technical issues have remained important throughout:[10]

1. The first is **reference levels**. If emissions are being reduced, it is important to know by how much. There is no easy way to determine this, as historical deforestation rates (assuming these are known, which is not always the case) are not necessarily a good guide to the future. Although satellite imagery is an increasingly reliable tool to monitor deforestation, the second D in REDD+, degradation, is still hard to monitor.

2. If deforestation was going down anyway, there would be no **additionality.** The dilemma on this issue is that if reference levels are too generous there is the risk of "hot air" and diluted incentives, but if they are set too tight there may be a lack of incentives for tropical countries. To complicate things further, there is a potential problem that countries with high deforestation rates can win rewards more easily than those protecting their forests well already.

3. Linked to this are the issues of **permanence**. If the reductions of defor-estation are not sustained, how should payments for performance be handled?

4. **Leakage** occurs if the protection of forests in one site simply means that clearance shifts elsewhere. Clearly this must be avoided if the reductions are to be real.

5. Three approaches have emerged concerning geographic **scale**. The first and most difficult is stand-alone projects. The second is subnational (state or provincial) and the third is national. Various com-binations are possible. Two things are clear, however: UNFCCC rules require national-level accounting, and the issues of leakage and addi-tionality are particularly challenging at the project level.

6. **Safeguards** to ensure that REDD+ does not have negative social or environmental consequences. There were significant concerns among indigenous peoples that their already tenuous rights over the forests they inhabited would be undermined by "carbon cowboys" acquiring and selling off the carbon rights to these same forests.

7. **Forest tenure** is inextricably linked to REDD. Tenure reform can sup-port REDD+ and REDD+ can provide an incentive to implement tenure reform, although both processes are difficult.[11]

8. There need to be systems in place to **monitor**, **report**, and **verify** progress.

   In the years between Bali and Warsaw, significant progress has been made on many of these technical issues, and this continues. It will be no surprise, however, that social issues continue to be challenging. One important breakthrough was recognizing that it would take time to set systems in place before international payments could flow and that the international community should provide financial support for this REDD+ readiness work. The phases of REDD+ were initially proposed in a report supported by the Norwegian government and the Climate and Land Use Alliance.[12]

Main players and funders

Norway emerged early on as a major supporter of REDD+, even though the country has not traditionally been a major player in tropical forest conservation. In 2007 Prime Minister Jens Stoltenberg launched the Norwegian Forest and Climate Initiative with a budget of USD 500 million per year. Norway went on to sign ground-breaking payments for performance agreements with the governments of Indonesia and Brazil for USD 1 billion each. At the time Stoltenberg said, "Through effective measures against deforestation, we can achieve large cuts in greenhouse gas emissions quickly and at low cost. The technology is well known and has been available for thousands of years. Everybody knows how not to cut down a tree."[13] Other major funders of REDD+ were Germany (which created a new International Climate Initiative partly to support REDD+), the United Kingdom, the United States, France, Australia, and Japan.

The major international conservation groups also became involved in REDD+, which they saw as a mechanism to support their ongoing forest conservation programs. Some, such as the Nature Conservancy and Conservation International, became actively involved in pilot projects. Others, such as WWF, focused more on the policy level. The nongovernmental organization community was divided about whether individual REDD+ projects should generate carbon credits that could be used to offset industrial emissions. Organizations of indigenous peoples, such as the Coordinator of Indigenous Organizations of the Amazon River Basin and the Southeast Asian Tebtebba, were actively involved in REDD+ negotiations at UNFCCC meetings, sometimes on their countries' delegations. Indigenous peoples were somewhat divided on whether REDD+ was going to be beneficial to them or a threat to their land rights. However, they have been effective in promoting safeguards in the UNFCCC context.

Among the tropical forest countries, Papua New Guinea initially played a leading role within the Coalition for Rainforest Nations. Brazil,

Indonesia, Vietnam, Peru, Gabon, and Mexico have emerged as the leaders in implementing REDD+ on the ground. Nongovernmental organizations such as the Amazon Environmental Research Institute (IPAM) and Imazon in Brazil, Kemitraan in Indonesia, and CCMSS in Mexico have played key roles. And the Center for International Forestry Research produced a series of books and reports to help policy makers and practitioners. The direct engagement of researchers in an emerging forest policy field is an unusual and encouraging development.

Progress to date: Opportunities and challenges

Looking back at what has been achieved, Arid Angelsen, a professor at the school of economics of the Norwegian University of Life Sciences and a world expert in the development of REDD+, noted in 2012, "REDD+ is moving ahead but at a slower pace and in a different form than we expected when we launched it in Bali in 2007 ... REDD+ as an idea is a success story. ... It also takes a fresh approach to the forest and climate debate, with large scale result-based funding as a key characteristic and the hope that transformational change will happen both in and beyond the forest sector." However, he also noted that reducing deforestation is difficult: "To fully realize its mitigation potential, REDD+ requires transformational change in the form of altered economic, regulatory, and governance frameworks ... a key factor for achieving transformational change lies in the autonomy of the state from key interests that drive deforestation and forest degradation and the presence of strong coalitions that call for such change to take place."[14]

BRAZIL AND INDONESIA: TWO CONTRASTING PICTURES OF PROGRESS ON REDD+

Deforestation in the Brazilian Amazon has decreased substantially from a high in 2004. One study suggests that approximately half this reduction can be attributed to policy initiatives by the Brazilian government and half to declining commodity prices (see also chapter 3).[15] In 2009 Brazil launched a national climate change policy that contained emissions reduction targets by sector, including a reduction in Amazonian deforestation of 80 percent by 2020. Although there was some leakage of deforestation into neighboring Amazonian countries and to the cerrado forests, Brazil's efforts have been remarkably successful. It should also be noted that although some foreign assistance has been provided, notably by Norway and Germany, the vast majority of costs have been covered by Brazil itself.

Indonesia's emissions from deforestation were second only to Brazil's in 2004, but the reductions in Brazil have not been mirrored in Indonesia. Unlike Brazil, Indonesia does not have reliable, publicly available information on deforestation. Conversion of Indonesia's extensive peatlands to palm oil or pulp and paper plantations are responsible for a significant part of these emissions (see chapter 3). Indonesia made a commitment to reduce its greenhouse gas emissions by 26 percent on its own, and increased its commitment to 41 percent with international assistance at the Copenhagen meeting in 2009. Most of these reductions would come from reducing deforestation. In May 2010 Norway signed a letter of intent with Indonesia for a process of preparation, REDD+ readiness, and contributions for verified emissions. The latter were to start in 2014 but will be delayed as progress has been slow. One key building block of the letter of intent was a moratorium on new clearing of natural forests. This was implemented by presidential decree in 2012 and renewed in 2013. Some forty-three million hectares of forest are covered by the moratorium, but forest users with existing licenses in the area are still permitted to operate. There are other exemptions, so the real impact is somewhat unclear. Another commitment concerned the creation of a "REDD+ agency" which has not yet been done either.

In hindsight it should not be a surprise that reducing deforestation is a little more complex than Prime Minister Stoltenberg's "not cutting down a tree." The global carbon market anticipated in Copenhagen has not materialized, nor has the United States adopted a cap-and-trade system providing a national market for carbon offsets. Deforestation is driven by powerful economic drivers, mainly agricultural but also increasingly mining and infrastructure. The scale of funding available for REDD+, although unprecedented, is several orders of magnitude smaller than funds being invested in these sectors. Local communities and indigenous peoples have difficulty in having their voices heard and their land rights protected in the face of these forces, and nongovernmental organizations may have more influence in national capitals than on the ground where forests are being cleared. However, it is too early to give up on REDD+, and many of the improvements in policies and practices that are being put into place as part of REDD+ readiness are beneficial in themselves. REDD+ may be moving more slowly than originally expected, but it still has the potential to be a game changer in countries with high deforestation rates.

CHRIS ELLIOTT is the executive director of the Climate and Land Use Alliance and is an adjunct professor in the Department of Forest Sciences at the University of British Columbia.

9

THE TWENTY-FIRST CENTURY
Decisive for the Tropical Rainforests!

THE GLOBAL CONCERN about tropical rainforests has gone through various stages over the past forty years at a roughly decadal pace: in the 1970s the first estimates of rapid deforestation shocked the world; in the 1980s intergovernmental agreements tried to address deforestation, ending with the World Summit in Rio de Janeiro; in the 1990s coalitions between nongovernmental organizations, bilaterals, financial institutions, and individual governments formed and forest remote sensing came to fruition.

The next metamorphosis was at the beginning of the new millennium, when tropical rainforests started to be seen as "carbon stock" that must be preserved to prevent greenhouse gas emissions. The result was a proposal to compensate countries for avoiding deforestation, an initiative called REDD (see Chris Elliott's Specialist's View on p. 235). Biodiversity had to take a backseat as apprehension about biodiversity loss and extinction rates gave way to concern about a single chemical element—carbon—and under this label transformed to become once more a multilateral negotiation issue.

Not surprisingly, in the mind of the person on the street the tropical rainforest is out of sight and therefore out of mind. This public perception problem is made worse when some specialists publicly proclaim that tropical rainforests only persist in a hand-ful of reserves, or consist only of degraded secondary forests. They declare an ailing patient in need of help to be dead already. I would be the last to underestimate the immense pressure on these forests and the real risk that such pessimistic predictions may come true by 2050, but ill-informed pronouncements do nothing to improve our chances of saving the remaining tropical rainforests.

Although the situation is grave, we should recognize that the most sophisticated assessment based on global satellite data esti-mated the tropical rainforest areas remaining in 2012 at over a billion hectares. An incomplete FAO assessment has estimated the primary forest areas remaining in 2010 at over 700 million hectares (see table 7.2)—an area the size of Australia—of which about one-third comes under some form of protection. Depending on how we account for primary versus degraded forests, it is fair to say that per-haps half of the tropical rainforest existing in historical times has disappeared. Should we forget about the other half, discard it as val-ueless simply because we are overwhelmed by the bad news?

Having revisited and analyzed the major subject areas of rel-evance for the future of these forests, and the survival of the immense biodiversity they still harbor, in this final chapter I will discuss the likely situation by the middle of this century—less than forty years from now. And I will suggest how we can keep the situ-ation from developing into a catastrophe by the end of the century. Even though I have tried to look at the facts in as cool and rational a manner as possible, I accept the risk that I may have overlooked important aspects. I therefore start with a word of caution:

1) This is not about predicting, or guessing, whether tropical rainforests will still exist by the middle of the twenty-first century or what they will look like then. "Prediction is very difficult, especially about the future," noted the Danish physicist Niels Bohr (1885–1962).

2) I cannot even claim it to be a forecast, which is a calculated future based on a rational assessment of all relevant parameters. Social and political parameters are notoriously difficult to forecast, and other parameters are not independent from each other. Consider how emissions from deforestation influence climate change, which in turn influences forest structure and distribution.

3) The only possible thing to do is to look at a number of trends and project their probable outcome under different scenarios. The difficulty with some of these trends is that they never follow a linear trajectory. For example, the deforestation trends in Malaysia of the 1980s were projected to result in a completely treeless country by 2000! Linear projections, on the other hand, indicate the rate of change without taking mitigating or enhancing parameters into account.

Global deforestation trends: Do they foretell the future?

A HUGE AMOUNT of effort has gone into assessing the area of tropical rainforests in the world. These efforts have been marked by problems with ill-defined forest classification and inconsistent geographic areas. Gross forest loss and net loss (which takes reforestation into account) have also been confused. In extreme cases these errors have caused a variance of more than 10 percent in area estimates. Deforestation rates calculated with these numbers could easily be turned from a loss to a gain, or vice versa. Only since multi-temporal satellite imagery has been available, allowing us to compare forest area pixel by pixel, do we understand deforestation dynamics with reasonable accuracy. Tropical rainforest area estimates are still important, but total area estimates and estimates of deforestation rates are not enough to describe the dynamics of change. The emphasis has partly shifted to examining the fragmentation, degradation, and contiguousness (connectedness) of forest areas. These may be more relevant criteria for the conservation of biodiversity, and its resilience to climate change, than total forest area assessments.

Having said this, it is important to recognize that the more reliable area assessments have clearly shown a decline in deforestation rates in the past few decades. The earliest systematic deforestation estimate by the FAO was Adrian Sommer's informed guess of eleven million hectares per year for the 1970s. If this was anywhere near the actual loss—we will never be able to verify this—the deforestation rate would have been halved in the past forty years. This decreasing trend has been confirmed for the period since the 1990s when deforestation rates could be more reliably assessed, and it holds for all tropical rainforest regions.

This prompts the following question: Is there a realistic hope that deforestation rates will soon reach a level of zero deforestation? Some governments as well as nongovernmental organizations have established targets of zero *net* deforestation in five or ten years, meaning that the area deforested would not be more than the area of forest gain.

ZERO NET DEFORESTATION IN THE FUTURE

THERE ARE SEVERAL important considerations for trying to answer the question of whether we, or our children, will see an end of deforestation and a transition to gradual forest recovery on a global scale.

1) Between 2000 and 2012 the world has been losing about 0.43 percent of its remaining tropical rainforests per year: about 1.2 times the size of Switzerland (see table 3.1 and figure 3.2). If the decline in

the global deforestation rate were to continue in a linear fashion, it would reach the level of current reforestation (about 0.13 percent) a few years after 2030. If reforestation rates would increase in the future, which may be possible because there is more forest fallow in the process of regrowth, the zero net deforestation point would be reached earlier.

2) Zero net deforestation, however, does not mean there would not be any deforestation when this point is reached; it would only mean that deforested areas would be replaced by reforestation (natural regrowth and tree plantations). Primary forest could still be lost. Since 2000, primary rainforests have been declining faster than the total forest area (see table 7.2 and chapter 7). If this rate (approximately 0.5 percent per year) remains the same over the next decades, a further primary rainforest area the size of France, Spain, and Italy together will be lost by the middle of the century.

3) Linear projections, as I have mentioned earlier, are not forecasts. They only indicate the speed of change. A rough assessment of this kind nevertheless raises an important point: net loss or gain of forest area refers to forest area without saying anything about its intactness. If, in the course of a transition from net deforestation to net gain, intact forests are replaced by fragmented lower-quality regrowth and timber plantations, the forest biodiversity will suffer greatly even though forest area is maintained. Thus, although maintaining as much forest area as possible is a laudable goal, for biodiversity conservation it may be a phantasm. This also applies to the carbon storage capacity, which is a function not only of area, but of forest structure and biomass as well.

We should therefore not get "area fixation," attributing too much importance to numbers of hectares. In order to ensure that biodiversity does not get lost, zero net deforestation targets need to be combined with zero degradation of intact forests. Some tropical rainforest countries will certainly be able to reach a zero net deforestation level following the examples of El Salvador, Costa Rica, or Vietnam. These will more likely be smaller countries with functioning forest governance systems and low levels of illegal logging, where forest conservation makes economic sense. However, the deforestation trends of the past two decades do not foreshadow a future without deforestation, much less a future without rainforest biodiversity loss. The reality, unfortunately, looks much grimmer. To assess that reality we need to take a deeper look into the nature of some of the most important parameters of change.

The dynamics of the coming decades

THERE ARE FOUR main parameters that will determine the future of the tropical rainforests:
1) Population pressure
2) Climate change
3) Commercial pressure
4) Governance

The first two of these, population pressure and climate change, are to a large extent predetermined by historical trends, whereas the other two, commercial pressure and governance, are socioeconomic and political variables. Each tropical region has its unique paleoclimatic history, ecology, and exposure to human influence, so these four parameters will affect them in different ways. And most tropical rainforest countries will not just be exposed to one of these parameters, but will suffer assault from multiple quarters.

POPULATION PRESSURE

RURAL POPULATION DENSITY has often been taken as the main parameter influencing deforestation rates. A few decades ago, when slash-and-burn farming by smallholders was still the main driver of deforestation, this relationship may have held true, but this has not been the case since conversion for large-scale commercial agriculture became the main deforestation driver. Twenty-five years ago the FAO used human population density to adjust forest cover data for FRA 1990, but their projections were shown to be inconsistent and abandoned soon thereafter.[1] Some authors have tried to model deforestation rates and future tropical forest cover as a function of declining birth rates and shrinking rural population numbers. They fostered hope that urbanization and land abandonment could slow down deforestation still further and lead to accelerated reforestation and forest transitions,[2] but urbanization has convincingly been shown to be itself a root cause of deforestation.[3]

As deforestation is now to a large extent caused by enterprise-driven commercial agriculture supplying commodities to a rapidly growing urban middle class, the correlation between rural population levels and deforestation is even further off the mark. Human population growth, whether rural or urban, translates into deforestation drivers in many ways, depending on the socioeconomic and political circumstances, even within individual countries. In some Central American countries, for example, the forest cover and regrowth correlate not with human population density, but with remittances from workers abroad.[4]

Population density is certainly not appropriate as a single indicator, but it is far from irrelevant. In the next thirty-five years population growth is projected to peak everywhere except in Africa, where the population will double from 1.17 billion in 2015 to 2.39 billion in 2052 and is expected to rise to 4.2 billion in 2100. Western, central, and eastern African countries show some of the highest growth rates at 2–3 percent (see figure 9.1). Without any doubt, this will put the African rainforests under great pressure as large areas of tropical rainforests will be cleared and cultivated by smallholders and mixed production farmers.

FIG. 9.1. Human population in billions since 1950 projected to 2100

Source: United Nations Department of Economic and Social Affairs, Population Division, 2012.

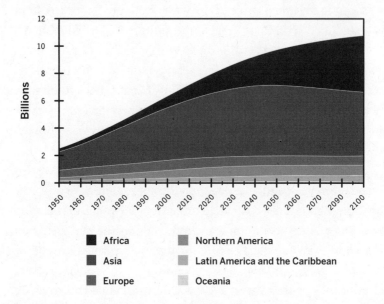

CLIMATE CHANGE: TIPPING POINTS AHEAD

IT IS, UNFORTUNATELY, realistic to assume that global warming will exceed 2°C above the pre-industrial average sometime after the middle of this century. The fifth IPCC assessment report shows a global mean temperature rise averaged across all medium emission scenarios likely to exceed the 2°C threshold by the end of the century, and a warming of 3°C–4°C under high emission scenarios. The lethargy of the international community, especially of those countries with the largest per capita emissions, toward effectively addressing dangerous climate change makes other scenarios unlikely. According to independent warming scenarios, the 2°C threshold may even be reached as early as 2050.[5] What does this mean for the future of the tropical rainforests?

The IPCC Working Group II report on the effects of climate change[6] relies on the research focused on Amazonia described in chapter 8. The report recognizes that higher temperatures will cause changes not by directly affecting rainforest plant physiology but by causing droughts and fires that push the forest across ecological thresholds. According to the working group's analysis, climate change without land use changes and fires would not drive large-scale forest loss in the twenty-first century, although there would be some shifts to drier forest types in the eastern Amazon. The seemingly strong resilience of tropical rainforests, however, presupposes that there will not be extreme droughts at the same frequency as occurred in Amazonia between 1998 and 2010. The analysis also assumes that deforestation is contained and that fires are controlled and extinguished—currently a rather optimistic scenario.

The future dynamic of the world's tropical forest ecosystems will be determined by a dangerous combination of climate change–induced droughts and fires on the one hand, and forest fragmentation and degradation on the other. In the coming decades the effects of climate change will be particularly severe in tropical rainforest areas that are exposed to fragmentation and forest degradation, whereas large contiguous, intact forest areas will be much less exposed. Forest fires across the humid tropics will increase in the coming decades and increasingly invade fragmented rainforest areas from the fringes. A gradual die-back of rainforest areas, particularly those receiving less than about 1,750 millimeters of rainfall in the Amazon, is therefore a realistic scenario, even in the first half of this century.

Africa, as a relatively dry continent, is also vulnerable to climate change effects. Its tropical rainforests still show aftereffects of the climate variations of the Pleistocene, and even though there are considerable climate model uncertainties, the continued impact of forest fragmentation by slash-and-burn farming exposes African rainforests to severe damage. In Southeast Asia, climate change effects may be less severe due to the higher rainfall patterns.

With a rise of global warming above the 2°C threshold, the possibility of tipping point effects—dangerous self-reinforcing climate change triggered by tropical rainforest die-back—cannot be ruled out. The Committee on Abrupt Climate Change of the US National Research Council concludes that a variety of climate change–induced effects make abrupt change in many tropical rainforest areas in the next fifty years plausible.[7]

COMMERCIAL PRESSURE: DRIVEN BY URBANIZATION

THE CONVERSION OF tropical rainforests for commercial agriculture, primarily palm oil production, soybean cultivation, and cattle pastures, has been the main driver of deforestation in the past twenty-five years. Significantly, the largest reductions in deforestation rates took place in Brazil, when forest conversion for commercial use was slowed down. This example demonstrates how deforestation could be reduced in other countries—if the right policies were adopted.

Commercial pressure for agricultural land conversion is coupled with a weakening state influence; today the production and trading of commodities from rainforest areas is in the hands of powerful corporations and is governed by financial flows. Their markets are a far cry from the village stands where the products from small-scale farms used to be traded. The new markets are increasingly urban and create an almost insatiable demand for some of the commodities that are produced on former rainforest soils.

A major concern is that oil palm plantations will continue to eat up lowland rainforest areas in Southeast Asia and increasingly encroach on African and Latin American forests. The projected global production increase of 40 percent by 2020 approximately matches the extent of the Indonesian area under oil palm plantations in 2012. Improvements in productivity may moderate the land requirement to a certain degree, but the skyrocketing of world production shown in figure 4.5 leaves little doubt about the tendency in the coming decades. If the demand for biofuels continues to rise, it will further exacerbate the pressure on rainforest areas. The economic pressure to convert land for biofuels production is a function of petroleum prices.[8] Palm oil is already being used as fuel for the domestic transport sector in Indonesia. Given that palm oil is produced at the expense of former tropical rainforests, the carbon footprint of this biofuel will be much higher than that of petroleum.

Urbanization and agricultural exports will continue to exert strong pressure for rainforest conversion to agricultural land in all tropical regions.[9] Over 70 percent of the world's population is expected to live in urban areas by 2050, which will increase the demand for food, mainly meat and vegetable oil but also other products like coffee and cocoa. The projected demand for land by 2030 has been calculated to lie between 285 and 792 million hectares.[10] Additional cropland and grazing land make up only about a third of this; the rest is expected to be used for industrial forestry, urban expansion, mining, and other land uses. As the humid tropics are among the very few geographic zones that still contain large areas of land suitable for large-scale agricultural expansion, it would be naive to believe that the pressure to convert tropical rainforest will miraculously subside all of a sudden.

GOVERNANCE: DECISIVE POTENTIAL

THE TERM "GOVERNANCE" refers to any process of governing, including defining plans, taking action, and verifying performance, be it by government, public–private partnerships, or networks of business, multilateral organizations, nongovernmental organizations, and local community groups. Such networks play an extremely important role in securing the future of tropical rainforests. Since the early 1990s the increasing influence of large agricultural and pulp and paper conglomerates has weakened the influence of the state in many tropical countries, and land use decisions have increasingly been replaced by enterprise-driven development.[11] This trend has also increased "crony capitalism" among local politicians and influential family members and has fostered corruption and dysfunction in the democratic systems of a number of countries. These countries are much less likely to preserve their forests than those with functioning and transparent governance systems.[12] Multi-stakeholder governance has an important role to play and can be decisive for the future of the rainforests.

It is certainly not my intention to downplay the ultimate responsibility of governments or the role of multilateral agreements such as the Convention on Biological Diversity (see box 9.1). Nonetheless, the examples described in this book have shown that, beyond government, it is a country's economic, social, and political pattern that determines to a large degree whether the tropical rainforests have a chance to survive or not. The more open and democratic a country is, the better such governance networks function and the more important is their role in conservation, the mitigation of corporate abuse and the fight against corruption. The example of the impressive reduction of deforestation in Brazil since 2004 is proof of the effectiveness of such networks in conjunction with political will and forceful government interventions.

BOX 9.1   **Infinite hope: The Convention on Biological Diversity**

It is a legitimate question to ask why the role of the Convention on Biological Diversity (CBD) has not appeared in a more prominent place in
this book. Among all the multilateral environmental agreements, this is
the one that should presumably be most relevant to the conservation
of tropical rainforest biodiversity.

However, of the three goals of the CBD, only one is concerned with
the conservation of biodiversity. The other two are focused on the sustainable use of biodiversity and the sharing of benefits from the use of
genetic resources. The latter had been a primary focus of the CBD in
the first fifteen years since it was created in 1992. In fact, until recently
the CBD did not focus on the root causes of biodiversity loss and failed
to achieve the 2010 targets to reduce the rate of biodiversity loss
decided at the World Summit on Sustainable Development in Johannesburg in 2002.

The CBD has been constrained by its inability to support the implementation of the convention by the member states, and also by an
overload of meetings and internal bureaucracy. At the 2010 Conference
of the Parties in Nagoya, Japan, an ambitious new strategy for 2011–
2020 was adopted. The plan includes a goal to address the underlying
causes of biodiversity loss and consists of twenty points—the Aichi
targets. As part of this plan, governments agreed to increase the area
under legal protection to 17 percent by 2020. It is too early to judge
whether the CBD and its member countries will be able to live up to the
ambitions of the Aichi targets by then.

We must accept finite disappointment, but never lose infinite hope.
MARTIN LUTHER KING, JR.

In a globalized market economy, it falls to international conservation, development, and human rights nongovernmental organizations to promote sustainable practices, along or through supply chains. When the soybean cultivation and cattle pastures were expanding into the rainforests of Rondônia, Mato Grosso, and Pará in Brazil, a relatively small number of large, internationally active corporations were behind it. The finite number of well-known business players connecting the primary producer far off in the Amazon Basin with the consumer in São Paulo, North America, and Europe exposed them to consumer reactions. It was the risk of losing their reputation among well-informed consumers that made them move quickly.

What has now become known as "teleconnection"—consumer behavior in one part of the world affecting land use (in this case deforestation) in geographically distant areas and across the supply chain—has successfully been used to promote sustainability. In the case of soy and beef production in Brazil, and to a certain degree also pulp and paper and illegal logging in Indonesia, teleconnections have been used effectively. Where such links are more diffuse or consumers are less aware, consumer pressure is a much less effective tool for changing corporate behavior, and ultimately national policy. This is an important reason why teleconnections have not been effective with palm oil producers: the producers are more obscure, their customers are more dispersed, and their product is used mainly in markets with low environmental awareness.

BOX 9.2

The World Bank–WWF Alliance for Forest Conservation and Sustainable Use

BRUCE J. CABARLE

The World Bank–WWF alliance was a response to the global forest crisis—the increasing loss of forest biodiversity, goods, and services essential for sustainable development. This unusual combination of partners believed that they would have greater strength if they combined their efforts to reduce poverty and protect the environment. The alliance's goal was to significantly reduce the rate of forest loss and degradation worldwide. From 1998 to 2005, the alliance worked with governments, the private sector, and civil society toward increasing the area of protected forest, improving the protection of existing reserves, and increasing the production forests under certification.

During its life span, the alliance helped establish the following:
1) 56 million hectares of new protected areas,
2) improved management for 82 million hectares of protected areas, and
3) responsible management of 31 million hectares of commercial production forest.

The alliance partners also raised significant funding that facilitated the investment of USD 300 million in conservation projects in over fifty countries in pursuit of the above targets. The alliance provided funding and supported the involvement of governments and other institutions in the following projects:
- The Amazon Region Protected Areas program. The first ten-year phase protected 12 percent of the Brazilian Amazon and established a USD 220 million trust fund to support the ongoing management of this network. The scope of the program was equivalent to building the entire United States national parks system in only a decade.
- The 1999 Yaoundé and 2005 Brazzaville Heads of State Forest Summits. These landmark meetings resulted in cross-border cooperation in

the Congo Basin on forest conservation and responsible management (see box 7.2). Following the first summit in Yaoundé, 3.5 million hectares of new protected areas were established in the Congo Basin.

- Research that led to the development of a systematic approach for the detection, prevention, and suppression of illegal logging in Indonesia. This brought together a wide range of stakeholders and helped develop a constituency for change, resulting in a presidential decree to combat illegal logging.
- The World Bank revised its Operational Policy on Forests in 2002. It endorsed independent, transparent, performance-based certification and recommended protecting critical forest areas identified by the analysis and proof-of-concept testing done by the alliance.
- The WWF amended its forest management target in 2005, recognizing the following points: the need to address illegal logging; the importance of incremental improvements in forest management, especially by local communities; the need for improved governance; and the value of nontimber forest resources and environmental services.

The alliance clearly demonstrated that rainforest conservation demands new forms of cooperation. We need to bring together expertise and resources from both the public and private sectors to produce the triple dividend of conservation, poverty alleviation, and sustainable development.

BRUCE CABARLE *is the president of Concentric Sustainability Solutions* LLC *and a senior fellow at the National Council for Science and the Environment. He is also the first chairman of the Forest Stewardship Council and a former director of* WWF*'s Global Forest Program.*

The tropical rainforests by 2050

TAKING THE FOUR parameters of population pressure, climate change impacts, commercial pressure, and governance into account, what will be the likely outcome in the three main tropical rainforest regions by the middle of this century?

LATIN AMERICA

IT IS LIKELY that the declining trend of deforestation rates will continue in Latin America, and that some of the smaller countries will reach zero net deforestation, albeit at a low level of forest cover and with virtually no primary forest left outside protected areas. The overall regional deforestation trend, however, will depend on a continued effort to reduce deforestation and illegal logging in Brazil. The modifications to the Brazilian Forest Code add some uncertainty. In the Andean countries of Peru, Colombia, Bolivia, and Ecuador, deforestation will continue to come at the expense of primary forests and may in fact increase in the wake of expanding cattle ranching, mining, and petroleum exploitation in lowland primary rainforest areas. Distinct signs of climate change impacts will become visible in the southern half of the Amazon Basin as droughts and fires in fragmented forest areas become more frequent, and these events will increasingly modify the forest structure.

ASIA–PACIFIC

THE DEFORESTATION RATE in the tropical rainforests of the Asia–Pacific Region is also likely to decline further, but that rate starts from a much higher level of gross deforestation—close to 0.8 percent annually (see figure 3.2)—than in the other regions. The regional deforestation rate is strongly influenced by Indonesia, the largest country in the region, which has a declared policy to double its palm oil production between 2011 and 2020.

Net forest loss, which takes reforestation into account, is not a reliable indicator for the deforestation trend in this country, as the distinction between natural forest regrowth and tree plantations is not clear. In Malaysia and Indonesia practically no primary lowland rainforest outside protected areas will remain by 2050. Looking at the current pressure for land conversion in Papua New Guinea it is likely that its primary forests will also be greatly reduced. Thus, while the regional net deforestation rate will decrease further, that decrease will be slower than hoped for and will not reach zero net deforestation by 2050. The rainforests of this region, on the other hand, will be less affected by climate change than those in the other regions of the world, with the exception of the remaining natural forests on peat soil, which are vulnerable to fires in drought years.

AFRICA

THE SITUATION OF the African rainforests is in many ways the most perilous, and to a certain degree also the most unpredictable. All four parameters of forest change point to unfavorable conditions. Deforestation rates in Africa have been lower than in the other rain-forest regions of the world (see figure 3.2) and if the slowing trend since the 1990s were to continue, zero net deforestation could be reached in the next thirty-five years. However, Africa is likely to lose its place as the region with the lowest deforestation in the next decade, as the demand for additional agricultural land will increase dramatically.

The high population growth in Central Africa will drive defor-estation by slash-and-burn farming primarily from the northeast and along the Congo River and its tributaries into the Congo Basin forests of the Democratic Republic of the Congo, with very little if any government control. Recent research has found that defor-estation increases rapidly once human population density exceeds 8.5 people per square kilometer.[13] Urbanization is also expected to progress in Africa, but in this region this will not alleviate the pressure on the forest because the demand for fuelwood, charcoal, and bushmeat from urban centers also leaves a heavy footprint on forest areas (see chapter 4). The pressure on African rainforests caused by the rising population will be further compounded by increasing pressure from commercial agricultural expansion, mainly for palm oil plantations, as more companies from Asia and Europe move into Central African countries.

There is still great uncertainty about the effects of climate change on African rainforests, but models for the twenty-first century show an increasing risk of drought in a few decades.[14] This risk will be particularly acute in the fragmented forest areas on the fringes of the Congo Basin. For these reasons the deforestation rate in Africa is very likely to increase steadily, strongly influenced by the development in the Democratic Republic of the Congo, and by 2050 no end to deforestation is likely to be in sight.

CONCLUSION

TROPICAL RAINFOREST LOSS and degradation will continue until 2050 and beyond at rates between 0.3 and 0.4 percent net loss and therefore not reach a zero net deforestation level by 2050. The net loss by 2050 would still be between 11 and 14 percent of the total global area that existed in 2012 (see table 3.1). This corresponds to an area of between two-and-a-half and three times the size of Spain. The proportion of primary forest in this loss will be less than in the first decade of this century because of much lower deforestation for pasture expansion in Brazil. The degradation of primary forest to secondary forests, conversely, will continue at a high rate, as a function of the further expansion of selective logging, including illegal logging, into hitherto untouched forest areas. The additional reduction of the primary forest areas may be as high as 100 million hectares by 2050, about twice the size of Spain. From the point of

view of biodiversity conservation, this tendency is particularly worrying (see box 9.3).

Of course these projections to the middle of this century are
affected by many uncertainties, but they are based on actual trends
and some likely nonlinear developments. I am, however, convinced
that the greatest threat to the world's tropical rainforests over the
next decades is the "dual poison cocktail" of forest fragmentation and
degradation combined with drought and fires under a rapid global
warming scenario. The decisive period for the long-term future of
the rainforests and the majority of their species diversity will be the
second half of the century, when global warming is likely to exceed
2°C above the pre-industrial global average. It will be too late then to
avoid a dangerous tipping point of self-reinforcing climate change.
The earlier we take forceful measures to mitigate catastrophic climate change and to make tropical rainforests as resilient as possible,
the better the chances are that we will save at least some intact remnant areas of what used to be the world's vast and remote jungles
into the next century. The following outlines a number of strategic
directions and actions to prepare us for this colossal task.

BOX 9.3   The sixth mass extinction is underway!

Sixty-five and a half million years ago, the Chicxulub asteroid, a solid rock about 10 kilometers in diameter, collided with the Earth and left a huge crater on today's Yucatán Peninsula of Mexico, about 180 kilometers wide and 20 kilometers deep. The glowing fragments of the asteroid, ash, and toxic gases ejected into the atmosphere caused forest fires, extended darkness, global cooling, acid rain, and ocean acidification. The impact exterminated all non-avian dinosaurs and an estimated 75 percent of all other species on the planet.[15] The Chicxulub asteroid caused an instant atmospheric catastrophe and the fifth mass extinction, long before the advent of the hominids. Most biologists, and I am one of them, believe that the sixth mass extinction is in progress now.

When the *Washington Post* staff writer Joby Warrick published an article in 1998 under the title "Mass extinction underway, majority of biologists say," it triggered a huge wave of reactions. To date, the Mysterium.com website has accumulated over 300 links to authoritative reports and articles with evidence that a mass extinction is beginning.[16]

When Joby Warrick published his article, less than twenty years ago, most biologists still saw habitat destruction as the cause of the mass extinction, but after reviewing a good part of the scientific literature on the effect of climate change on tropical rainforest ecosystems, I have reached the following conclusion: climate change exacerbates habitat destruction to such a degree that by the end of this century—in the lifetime of our children and grandchildren—we will probably lose up to a third of the world's species (some scientists expect up to half of species to disappear).

The majority of these species will have vanished as their tropical rainforest habitats decline, victims of the "dual poison cocktail" of continued tropical rainforest loss and degradation on one hand, and drought and fires induced by climate change on the other. Another major part of the species loss will result from the climate change–induced disappearance of the majority of the world's coral reefs, and the rest will be due to a variety of ecosystem damages. This sixth mass extinction will be more gradual than the one caused by the Chicxulub asteroid, but only slightly so. It is being caused now, by one single species—*Homo sapiens*—yes, *sapiens*—knowingly.

Key messages for the future

THESE SEVENTEEN KEY messages are written with a focus on the conservation of intact tropical rainforest areas and their biodiversity as well as the survival of indigenous forest communities. I also recognize the importance of these forests for carbon storage and other ecosystem services, but priority actions to ensure these values may differ. These key messages are presented in four groups: principles of conservation, addressing root causes, strategic directions, and alliances and partnerships.

PRINCIPLES OF CONSERVATION

1) **Contain further resource exploitation in intact tropical rainforests**
   The value of the remaining intact, primary forests for biodiversity conservation as well as rainfall patterns and carbon storage should be better recognized. As timber production from tropical rainforests is likely to grow over the next decades, production should not lead to further expansion of logging activities into primary forest areas. Wood products should come from well-managed concessions and the large areas of existing secondary forests. Tree plantations for the production of timber, pulp and paper, rubber, and other products should never be established at the expense of primary forests.

2) **Avoid forest fragmentation and fringe effects**
   Forest fragmentation increases the risk of local species extinctions and exposes the remaining forests to such deleterious fringe effects as damage from extreme weather events, drought, fires, and over-hunting. As shifting agriculture by smallholders is a common cause of forest fragmentation, agroforestry methods that promote permanent agriculture and slow the progress of fragmentation should be introduced. Zoning, satellite surveillance of deforestation, and forest

law enforcement should be used to conserve contiguous rainforest areas. Special attention should be paid to the rapidly increasing forest fragmentation in Central Africa.

3) **Protect the cultures and legal rights of indigenous peoples**

There has been a rise in indigenous territories under official government protection in Latin America. These forest areas have often shown as much resistance to deforestation as other protected areas. The human and legal rights of indigenous communities should be guaranteed, and their defense against encroachment by immigrant farmers, illegal loggers, miners, and hunters should be supported. The REDD+ program should strengthen the rights of indigenous peoples and provide support for land demarcation, titling, and enforcement as an effective means to reduce carbon emissions, rather than limiting their rights over their land.

ADDRESSING ROOT CAUSES

4) **Prevent dangerous climate change**

Mitigating climate change is of the utmost importance for the long-term survival of the tropical rainforests, as much as it is crucial for the survival of marine, arctic, and other ecosystems. All other actions to save the extent and character of the tropical rainforests will be greatly impaired, if not rendered futile, without a substantial reduction of greenhouse gas emissions. There is very little time left to reduce the emissions from fossil fuel combustion, as 2°C of warming above the pre-industrial average is likely to be inevitable already.

5) **Keep food and feed out of the forest**

The footprint of food is the most serious threat to the integrity of tropical rainforests. Producing meat for human consumption by sacrificing tropical rainforest areas is the most inefficient and carbon-intensive use of forest land and should be completely avoided.

Investments in cattle and animal feed production in tropical rainforest areas by multilateral agencies must be stopped. Latin American countries should learn from the experiences of Brazil and not repeat past mistakes. Any further expansion of palm oil plantations should no longer come at the expense of intact forests, but be planned on existing areas of degraded former forest land. Investments in Africa by international agencies and corporations should follow the same principles.

6) **Reduce food waste in rapidly rising urban populations**
As food habits, particularly meat consumption, change with urbanization and rising income levels, one of the most effective resource preservation measures will be to reduce food waste. Cutting food waste in half over the next forty years could reduce the necessary agricultural production, and thus land requirement, to up to one-third less than the production needs currently projected by the FAO. The increasing food waste problem should particularly be addressed in the urban centers of the large emerging Asian economies.

7) **Ban biofuel production in tropical rainforest areas**
Producing biofuels at the expense of forest areas will increase global carbon emissions, and must be prevented under any circumstances. Since biofuel production is driven by petroleum prices, national legislation and international consumer standards need to regulate against biofuels produced in tropical rainforest areas, and the European Union and North American countries should introduce legislation banning biofuel imports from rainforest areas. Exceptions should only be made for local production of palm oil in poor rural communities from smallholder plantations.

8) **Combat illegal logging and reinforce international cooperation**
Illegal logging is a major contributor to deforestation and undermines the livelihood of local communities. It causes up to 22 billion

metric tons of carbon dioxide emissions annually.[17] Reducing emissions from deforestation therefore requires a forceful effort to curb illegal logging. Developing criminal justice strategies against illegal logging will require better domestic cooperation between enforcement agencies and civil society. This should be a component of development assistance programs. The voluntary partnership agreements with producer countries under the European Union FLEGT program should be accelerated, and their legality assurance systems should cover all harvesting, processing, and exports, not just those destined for Europe.

STRATEGIC DIRECTIONS

9) Expand protected tropical rainforest areas

Tropical rainforests are still underrepresented among the world's protected areas, particularly in Central Africa but also in other parts of the tropical world, for example in Indonesia, Papua New Guinea, and in some of the Andean countries. Protected areas should be designed to protect as much of the forest biodiversity as possible. To avoid the problem of "paper parks," the international community should support good management of protected areas. As pressure on intact forest areas continues to increase, there is not much time left to complete the protected areas system. The 2011–2020 strategic plan of the Convention on Biological Diversity with its Aichi targets could provide the framework for an increase in protected areas coverage, but requires substantial support from the international community and nongovernmental organizations.

10) Promote reforestation and recognize the value of secondary forests

The negative effects of forest fragmentation can be moderated by reconnecting the fragments, either by letting regrowth progress in intermediate areas or by reforestation through tree planting in surrounding areas. Secondary forests may develop quickly in the surrounding matrix vegetation if soils and seed stocks have not been depleted through overgrazing and repeated burning. Protecting intact forest remnants in fragmented landscapes is particularly important, as they may serve as residual areas for seeds, animal seed dispersers, and pollinators. Fragmented landscapes may recover to closed secondary forests in ten to twenty years if protected from fires and other disturbances.[18] The Global Partnership for Forest Restoration of IUCN could be a relevant partner organization.[19]

11) Promote sustainable forest management practices and certification

Sustainable forest management practices contribute to forest protection, maintain natural capital value, and ensure a lasting flow of ecosystem services. Currently less than a third of production forest in humid tropical zones is managed under an approved management plan and less than 7 percent is certified. FSC certification should be much more forcefully promoted, and timber originating from FSC certified forest areas should be recognized as fulfilling the EU Timber Regulation/FLEGT legality requirements. The FSC has achieved widespread uptake in temperate and boreal zones; it should not rest on its laurels. It should reduce its internal bureaucracy and put more effort into promoting certification in tropical forests.

12) Engage Asian consumer countries in commodity standards

Asian markets for timber, pulp and paper, palm oil, and other commodities from tropical rainforest countries are huge. More effort should be made to promote production and trade standards, such as FSC certification for timber and the Roundtable on Sustainable Palm Oil standards for palm oil, particularly in China and Japan. As the majority of illegally sourced wood imported by Western countries comes from Asian processing countries, we must pursue initiatives to clean up the supply chains. Internationally active non-governmental organizations should cooperate with Asian countries to introduce standards for public procurement, consumer information, and education programs.

13) Promote forest monitoring systems with remote sensing technology

Forest monitoring with satellite remote sensing has provided effective tools for forest cover assessments and revolutionized the possibilities for forest law enforcement. The Brazilian PRODES and DETER products clearly demonstrate how these methods can be used to curb deforestation. However, in most tropical rainforest countries these tools have either not been available or not been used effectively. The publicly available high-resolution global forest cover change map[20] has improved this situation significantly. International cooperation and support through the REDD+ program, intergovernmental agencies, such as the World Bank Group, and bilateral development agencies can help countries to build the necessary human capacity to interpret satellite imagery and enforce forest law.

14) Improve agricultural productivity in Africa

The pressure to produce more food will be particularly acute in Africa because of its high population growth. Large-scale commercial agriculture will not meet the food requirements of the rapidly growing rural communities, which will remain heavily dependent on local food production. A substantial increase in productivity per area will be necessary if massive deforestation is to be avoided. There are many good examples of agroforestry systems and measures to increase soil fertility to allow more long-term land use. Rehabilitation of degraded land for agricultural use and reforestation should be strongly encouraged. The productivity of such neglected crops as yams, plantains, and cassava could also be increased by the introduction of better genetic varieties.[21] Development should focus on these objectives and stay away from cash-intensive agricultural gigantomania.

ALLIANCES AND PARTNERSHIPS

15) Promote the positive role of the state

With the increasing influence of enterprises in land use change, the state is called upon to have a more active role in monitoring, controling, and enforcing forest laws. As the case of Brazil has shown impressively, with the right kind of policies deforestation rates can be slowed effectively. The trade-off between agricultural expansion and the conservation of tropical rainforests can be addressed through such policy interventions as spatial land planning, land use efficiency gains, agricultural intensification, protected areas, and sustainable forest management. Civil society organizations can play a crucial role in supporting state agencies in monitoring and enforcement.

16) Speed up REDD+ implementation, but don't expect miracles

The initial enthusiasm about REDD has been tempered by its slow progress and the realization that deforestation is a more complex phenomenon than initially assumed. Deforestation today is driven by powerful economic actors, mainly in agriculture but increasingly in mining and infrastructure as well. Funding generated by REDD+ will remain orders of magnitude smaller and without a reform of economic, regulatory, and governance frameworks, REDD+ will never be able to live up to the expectations. The REDD+ readiness programs, nevertheless, have the potential to improve policies and procedures and need to be pursued.

17) Build multi-stakeholder alliances

Since the 1992 Earth Summit in Rio, the struggle to safeguard the tropical rainforests has shown the power of alliances between unequal and at times unconventional partners working on a common goal. The examples of the Yaoundé Process and the Amazon Region Protected Areas plan demonstrated that a cooperation between national governments, intergovernmental organizations, financial institutions including the Global Environmental Facility, local and international nongovernmental organizations, and business groups can be very effective in setting agendas and harnessing political and financial support for the conservation and sustainable use of tropical forests. The role of initiating and orchestrating such alliances falls to the better connected international nongovernmental organizations, as they are often perceived to be less embroiled in institutional politics. Such partnership networks enhance peer pressure and may help stop corruption and illegal activities that governments are sometimes unable to tackle alone.

A final word

I CANNOT REMEMBER having ever met a person, be it a scientist, corporate manager, politician, or indigenous person, who felt tropical rainforests should be cleared. Everybody asserts how important their protection is. But declarations, unfortunately, do not change the realities. Tropical deforestation is a highly complex process with different root causes in different parts of the world. When I started writing this book, I had seen many rainforests on all continents and I had worked to protect them during my entire professional life, as a scientist, national parks manager, and CEO of WWF International, one of the largest international conservation organizations, which made forest conservation one of its main priorities. I had helped orchestrate some of the alliances such as the Amazon Region

Protected Areas plan and the Yaoundé Process described above. Thus, I had no illusions about the dire prospects for many tropical rainforests and the complexity of deforestation processes.

The deeper I got into my analysis, the more I realized how much the scientific information on tropical rainforests had expanded in the past few years, and how much of this evidence now documents their gradual demise and that of their biodiversity. It felt indeed like the "premonition that paves the way to later knowledge" that Alexander von Humboldt described. But this evidence is so widely dispersed across many books, dozens of scientific journals, and specialized websites that it does not coalesce into a coherent call to action on the radar of the international community.

The most disturbing discovery I made had nothing to do with the many individual deforestation drivers and the impotence of the multilateral system in dealing with them—I knew those well enough. No: it was learning that the fatal interactions between these threats amplify the risks and could mean the almost complete disappearance of intact tropical rainforests in the twenty-first century.

Tropical rainforest conservation, the preservation of the world's biodiversity, and climate change mitigation are intrinsically linked and no single measure, tool, or policy will fix this fatal conundrum. There is no silver bullet to solve global environmental problems in a time of a globalized economy and weak governance systems. I am not a fatalist and I abhor statements that pretend all is lost—they are wrong. But without concerted action along the path I have outlined above, we may indeed lose it all. What we should never forget is that we humans are also a part of our world's biodiversity, and the security of our descendants will not be independent from it.

APPENDIX 1

COMPARISON OF TROPICAL moist forest (TMF) climax areas and estimates for the late 1970s in millions of hectares. Area estimates concern the sixty-five countries with tropical rainforest included by Sommer (1976).[1] The FAO/UNEP (1981–82) figures are likely to include some tropical dry forests. Both assessments, however, underestimated the actual tropical rainforest area as it existed in the 1970s.

REGION (NUMBER OF COUNTRIES AS IN SOMMER, 1976)	APPROXIMATE TMF CLIMAX	TMF (SOMMER, 1976)	TMF (FAO/UNEP 1981–82)
East Africa (6)	25	7	13.2
Central Africa (6)	269	149	170.3
West Africa (11)	68	19	21.0
TOTAL AFRICA	**362**	**175**	**204.5**
South America (11)	750	472	585.8
Central America and the Caribbean (13)	53	34	49.0
TOTAL LATIN AMERICA	**803**	**506**	**634.8**
Pacific (6)	48	36	33.7
Southeast Asia (9)	302	187	202.8
South Asia (3)	85	31	49.2
TOTAL ASIA	**435**	**254**	**285.7**
TOTAL WORLD	**1,600**	**935**	**1,125.0**

APPENDIX 2

Satellites and sensors in past and
current use for forest cover mapping

BASED ON INFORMATION from the Satellite Imaging Corporation,[1] Fuller (2006),[2] Achard and Hansen (2012),[3] and Wikipedia.[4]

1. Passive sensors

PASSIVE SENSORS REGISTER optical and thermal radiation. They can distinguish between land cover types in high resolution, but do not penetrate cloud cover. There are four types of resolution:
· Spatial resolution is the area on the ground covered by one pixel of the image.
· Spectral resolution refers to a sensor's ability to distinguish between electromagnetic wavelengths. Sensors may be panchromatic (shades of gray), or multispectral (uses several wavelength bands to produce imagery in color).
· Temporal resolution: the time between imagery collection (e.g., days).
· Radiometric resolution: the ability to record levels of contrast.

HERE IS A list of the most common passive sensors, categorized by spatial resolution:

a) **Low (coarse) spatial resolution optical sensors**
   (250 × 250 meters to 1 × 1 kilometer)
   - AVHRR (advanced very high resolution radiometer): This sensor maps on the basis of coarse spatial resolution from National Oceanographic and Atmospheric Administration (NOAA), with near daily coverage and resolution of 1.1 kilometers.
   - MODIS (moderate resolution imaging spectroradiometer): This is a sensor on the Terra and Aqua satellites. Coarse spatial resolution (250 meters, 500 meters, and 1 kilometer) data that covers the entire Earth almost every day—best for cloud-free images.
   - MERIS (medium resolution imaging spectrometer): This sensor is on the Envisat satellite and maps at a 300-meter resolution. It can be used for global wall-to-wall mapping, produced through the European Space Agency's GlobCover initiative.

b) **Medium (moderate) spatial resolution optical sensors**
   (20 × 20 meters to 100 × 100 meters)
   - **Landsat 5**, launched in 1984 and decommissioned in 2013, carried the Multispectral Scanner with sixty-meter resolution, initially included on Landsat 1 to 4, and the Thematic Mapper with thirty-meter resolution.
   - **Landsat 7** was launched in 1999. It carries the Enhanced Thematic Mapper Plus (ETM+) sensor, which has thirty-meter resolution and six optical bands. In 2003, a failure in the scan line corrector resulted in data gaps.
   - **Landsat 8** was launched in February 2013 with the Operational Land Imager sensor (thirty-meter multispectral spatial resolution) and the Thermal Infrared Sensor.[5]
   - SPOT (système probatoire pour l'observation de la terre): SPOT 2, 4, and 5 provide images with resolution from 2.5 meters to 1 kilometer. SPOT satellites can be used to acquire images of any place on the planet every day and at various resolutions.

- The **Sentinel** satellite series is part of the Copernicus program of the European Space Agency. It will include SAR (synthetic aperture radar) and optical sensors specifically aimed at vegetation monitoring and biomass estimation at a high revisiting frequency for cloud-free imagery.
- **CBERS-1** and **2** satellite–sensor combinations were launched by the China–Brazil Earth Resources Satellite Program in 1999 and 2003 respectively. Their resolutions range from 20 to 260 meters, and they are used to control deforestation and fire in the Amazon, as well as for other applications.

c) **High spatial resolution (0.4 × 0.4 meters to 4 × 4 meters) optical sensors**

THE FOLLOWING SATELLITES carry high-resolution sensors that are mostly commercially exploited and not suited for larger-scale forest cover mapping:

- **Ikonos** was launched in 1999. It is commercially operated by Geo-Eye and has 1- to 3.2-meter spatial resolution. Mainly used for urban and industrial purposes or vegetation on a local scale.
- **GeoEye 1** was launched in September 2008. It has the highest resolution of any commercial satellite: 0.41 meters in panchromatic mode and 1.65 meters in multispectral mode.
- **QuickBird**, owned by DigitalGlobe, is used to study small areas for things like insect damage to trees. It is impractical for larger areas.

2. Active sensors

ACTIVE SENSORS EMIT radiation and register reflection, using synthetic aperture radar (SAR). They can penetrate cloud cover and can produce three-dimensional pictures of the surface.

- **JERS** (Japanese Earth Resources Satellite) produces cloud-penetrating SAR imagery. JERS-1 measures canopy texture at 100-meter resolution. Used in the Global Rain Forest Mapping Project.

- ERS-1 and ERS-2 (European Remote Sensing Satellite) provide cloud-penetrating SAR imagery. ERS-2 is used to locate burning sites.
- **Lidar** (light detection and ranging laser) uses different wavelengths of laser to measure the depth of objects. It can be used to assess canopy height and biomass.
- BIOMASS is a long-wave SAR scheduled to be launched by the European Space Agency in 2020 to measure forest biomass.
- The DESDynI satellite will be launched in 2019. It will integrate SAR and lidar technology to produce accurate measurements of ecosystem structure and forest height.

APPENDIX 3

Forest classes used by the TREES
project and global net humid tropical
forest cover change 1990–1997

TWO PARAMETERS ARE used to define forest classes (see table below):

1) Forest proportion, that is, the percentage of forest within a mapping unit. Areas with a forest proportion of less than 40 percent were considered to be nonforest, with forest mosaics having 10–40 percent forest.

2) Tree cover, that is, the crown cover or canopy density. The minimum tree cover limits of 10 percent for an open forest and 40 percent for a closed forest are applied. Plantations and forest regrowth were included in the forest classes as nonnatural forest. For the calculation of forest cover, forest classes were weighted depending on the forest proportion per sample unit.

FOREST CLASS (FOREST COVER WEIGHT FOR CALCULATIONS)	FOREST PROPORTION (WITHIN SAMPLE UNIT)	TREE COVER (CANOPY DENSITY)	
		10%–40%	> 40%
Nonforest (25)	10%–40%	Forest mosaics	
Forest (75)	40%–70%	Fragmented forest	
Forest (100)	> 70%	Open forest	Closed forest

THE TOTAL FOREST cover estimates for 1990 and 1997 are the sums of the weighted forest cover areas in each forest class (see table below). The net decrease of thirty-four million hectares corresponds to a net annual change of -4.9 ± 1.3 million hectares. As deforestation as well as reforestation and natural regrowth takes place in all forest classes, actual deforestation rates cannot be directly computed from these net figures.

FOREST CLASSES	FOREST COVER IN 1990 (MILLION HA)	FOREST COVER IN 1997 (MILLION HA)	NET CHANGE 1990–1997 (MILLION HA)	ANNUAL NET CHANGE (%)
Closed	944	908	−36	−0.54
Open	130	134	+4	0.44
Fragmented	36	34	−2	−0.79
Plantations/ regrowth	9	9	0	0
Nonforest classes (mosaics)	31	31	0	0
TOTAL NET COVER	1,150	1,116	−34	−0.43

Information adapted from Achard et al. (2002).[1]

APPENDIX 4

Forest areas and net
change from FRA 2010

REGIONAL FOREST AREA trends in million hectares and annual change in percent for 1990–2000, 2000–2005, and 2005–2010 from the country tables in FRA 2010, as well as for a selection of countries falling entirely or predominantly in the humid tropical forest (HTF) distribution area. In this selection South and Southeast Asia includes Bangladesh, Sri Lanka, Brunei, Indonesia, Malaysia, the Philippines, Vietnam, and Papua New Guinea (PNG). West and Central Africa includes Ghana, Guinea-Bissau, Côte d'Ivoire, Liberia, Sierra Leone, Togo, Cameroon, the Republic of the Congo, Equatorial Guinea, Gabon, and the Democratic Republic of the Congo. South and Central America and the Caribbean includes Brazil, Colombia, Ecuador, French Guiana, Guyana, Peru, Suriname, Belize, Costa Rica, Cuba, the Dominican Republic, El Salvador, Guatemala, Haiti, Honduras, Jamaica, Nicaragua, Panama, and Trinidad and Tobago.

REGION	1990	2000	2005	2010
SOUTH AND S.E. ASIA (INCL. PNG) (MILLION HA)	356.9	331.3	328.8	323.1
Annual change for entire region (%)		-0.72	-0.15	-0.34
Annual change for HTF countries (%)		-0.96	-0.17	-0.43
WEST AND CENTRAL AFRICA (MILLION HA)	359.8	343.43	335.8	328.1
Annual change for entire region (%)		-0.46	-0.45	-0.46
Annual change for HTF countries (%)		-029	-0.30	-0.30
SOUTH AND CENTRAL AMERICA + CARIBBEAN (MILLION HA)	978.1	932.7	909.7	890.8
Annual change for entire region (%)		-0.46	-0.49	-0.42
Annual change for HTF countries (%)		-0.46	-0.49	-0.39

GLOSSARY

afforestation: Planting of forests on nonforest land (see also box 2.2).

agroforestry: A combination of trees and agricultural crops that helps protect soil and water for more sustainable land use.

ARPA: Amazon Region Protected Areas Program of Brazil.

ATIBT: Association Technique Internationale des Bois Tropicaux.

AVHRR: Advanced very high resolution radiometer (see Appendix 2 on satellite sensors).

biodiversity: Also called "biological diversity," this is the variation of life, consisting of the genetic diversity, the diversity of species, and the diversity of ecosystems in a given area.

biofuels: Fuels produced from living organisms (biomass), commonly as bioethanol or biodiesel.

biogeographic domain: Areas characterized by specific types of species and ecosystems.

biogeographic realm: Large geographic region (usually part of a continent) where ecosystems share a common evolutionary history, for example the Afrotropical realm or the Indo-Pacific realm (sometimes referred to as an ecozone).

biomass: Biological material from living or dead organisms. It most often refers to plants or plant-based materials.

biome: A major ecological community of plants and animals adapted to specific environmental conditions, for example desert, savanna, humid tropical forest, or freshwater swamp.

carbon sink: A natural or artificial reservoir of carbon. A forest is a natural reservoir of carbon through the sequestration of carbon dioxide from the atmosphere.

CBD: United Nations Convention on Biological Diversity (see box 9.1).

cerrado: Wooded grassland (savanna) area south and southeast of the Amazon Basin.

Chatham House: The Royal Institute of International Affairs, which leads in independent analysis and policy debate.

CI: Conservation International, a United States–based conservation organization, active in thirty countries.

CIFOR: Center for International Forestry Research, based in Bogor, Indonesia.

climate flywheel: Atmospheric circulation of humid air masses.

climax vegetation: Late succession stage (old growth) of a vegetation type. Climate (temperature, rainfall, seasonality), soil type, and other factors determine the climax vegetation.

closed forest: Forest area with a tree crown density of more than 40 percent (see appendix 3).

Club of Rome: A global think tank founded by Aurelio Peccei and other world leaders in 1968. It became world renowned with the publication of *The Limits to Growth* in 1972.

COMIFAC: Commission des Forêts d'Afrique Centrale (Central African Forest Commission) (see box 7.2).

commercial agriculture: Agriculture other than subsistence agriculture, usually dominated by large companies.

confidence limit: The upper and the lower limits of the confidence interval for a mean value. A 95 percent confidence interval contains 95 percent of all values.

COP: Conference of the Parties of International Conventions.

deep ecology: An environmental philosophy that recognizes the inherent worth of all living beings and considers humans as no more important than other species.

deforestation: See box 2.2.

DETER: A satellite-based system for real-time detection of deforestation in the Brazilian Amazon.

drought deciduous: Plants that drop their leaves during dry seasons or drought periods.

ecoregion: An ecologically and geographically distinct subunit of a biome, for example the "central Congolian lowland forests."

ecosystem services: Benefits provided by ecosystems, such as clean drinking water, medicinal plants, soil protection, and rainfall patterns.

ecozone: a large biogeographic area, for example the Neotropical or the Indo-Malayan ecozone.

El Niño–Southern Oscillation (ENSO): Periodic warming of ocean surface water off the Pacific coast of South America (El Niño), coupled with an oscillation in surface air pressure between the eastern and western Pacific Ocean.

emergent tree: A tree whose crown emerges above the crown canopy. It may reach heights of 50–60 meters.

endemic species: Plant or animal species that are native and occur only in specific areas or regions. For example all lemurs (a group of primates) are endemic to Madagascar: they occur only there.

ENSO: See El Niño–Southern Oscillation.

EROS–Landsat archive of the US Geological Survey: See appendix 2.

EUTR: EU Timber Regulation. This came into force in March 2013 and prohibits placing illegally harvested timber on the European Union market.

evapotranspiration: The sum of evaporation and plant transpiration of water to the atmosphere.

externalization: Displacement of (environmental) costs outside a system, usually a country.

fallow or forest fallow: Woody vegetation developing in abandoned clearings in areas of shifting agriculture.

FAO: United Nations Food and Agriculture Organization.

FAOSTAT: Data platform of the FAO Statistics Division.

FLEGT: Forest Law Enforcement, Governance and Trade, an action plan of the European Union to combat illegal logging.

Forest Code of Brazil: Legislation introduced in 1965 that required landowners to keep up to 80 percent of their land under forest. The code was revised in 2012, which created worldwide protests. It was challenged by Brazil's constitutional defender in 2014.

forest degradation: See box 2.2.

forest fragmentation: See box 2.2.

forest transition: Gradual recovery of deforested and degraded forest areas (see chapter 5).

FRA: Forest Resources Assessment of the FAO.

FSC: Forest Stewardship Council (see box 4.1).

FUNAI: The Brazilian National Indian Foundation.

GEF: Global Environmental Facility, a partnership for international cooperation to address global environmental issues established in 1991 by the United Nations Development Programme, the United Nations Environment Program, and the World Bank.

GPS: Global Positioning System.

GRID-Arendal: An environmental data center established by the government of Norway in collaboration with UNEP.

Holocene: Latest epoch of the Quaternary Period, from 11,700 years before present to the present.

humid tropical forest: See box 1.1.

IFC: International Finance Corporation of the World Bank Group.

igapó: "Swamp forest," seasonally flooded blackwater forest, on low-nutrient soil along the Rio Negro in the Amazon Basin.

IISD: International Institute for Sustainable Development.

indigenous territories (reserves): Areas inhabited and legally possessed by indigenous peoples.

INPE: Brazilian National Institute for Space Research.

intertropical convergence zone: A low-pressure trough that spans the globe near the equator with a band of clouds and thunderstorms. Over land, it influences rainfall patterns as it seasonally moves back and forth across the equator.

IPCC: Intergovernmental Panel on Climate Change.

ITTO: International Tropical Timber Organization.

IUCN: International Union for Conservation of Nature (also World Conservation Union). Founded in 1948 with 1,200 nongovernmental and governmental member organizations.

JRC: European Commission Joint Research Centre.

Lacey Act: US legislation that prohibits trafficking in illegally harvested species.

Landsat: See appendix 2 on satellite sensors.

lidar: See appendix 2 on satellite sensors.

LPI (Living Planet Index): A measure of the world's biological diversity developed by the WWF and the Zoological Society of London.

matrix vegetation: Structure and composition of surrounding vegetation.

MEA: Millennium Ecosystem Assessment of the United Nations.

MODIS: See appendix 2 on satellite sensors.

Mongabay: Online magazine in many languages, founded by Rhett A. Butler in 1999. Initially it focused on tropical rainforests, but it now also covers other environmental topics.

NASA: US National Aeronautics and Space Administration.

OECD: Organisation for Economic Co-operation and Development, currently with a membership of thirty-four (mainly industrialized) countries.

open forest: Forest area with a tree crown density usually between 10 and 40 percent (see appendix 3).

paleoclimates: Climatic conditions as they prevailed in earlier periods of the Earth's history. Paleoclimatology uses ice core and sedimentary (pollen) analysis methods.

palynology: The study of microscopic sedimentation. Often more narrowly used for the study of pollen and spores.

paper parks: A term used to describe protected areas that lack effective protection and management systems and therefore exist only on paper.

peatland: Areas of forest growing on peat (thick soil layers of dead or decaying plant matter). Peat forests are particularly common in lowland areas of Kalimantan.

planted forest: See box 2.2.

Pleistocene: Early epoch in the Quaternary Period, spanning from 1.8 million years before present to the beginning of the Holocene, 11,700 years before present.

primary forest: See box 2.2.

PRODES: Amazon Deforestation Monitoring Project of the Brazilian Space Research Institute (INPE).

Quaternary Period: Most recent period in the Earth's history (from 1.8 million years before present to the present). It consists of the Pleistocene followed by the Holocene.

Rainforest Action Network: A nongovernmental organization founded in California in 1985 known for its "environmentalism with teeth" campaigns against damaging corporate interests.

Rainforest Alliance: A nongovernmental organization founded in 1987 and based in New York. It uses market forces to promote sustainable use with its own certification scheme and characteristic frog logo.

Rainforest Foundation: A nongovernmental organization founded in the United States in 1989 by Sting and Trudie Styler mainly to protect indigenous peoples' rights, with organizations in the United Kingdom and Norway.

REDD: Stands for "Reducing Emissions from Deforestation and Forest Degradation." REDD+ includes such measures as conservation, sustainable forest management, and increase of carbon stock.

reforestation: See box 2.2.

refugia: Forest areas believed to have persisted as tropical rainforests during the climate fluctuations (glacial periods in the northern hemisphere) of the Pleistocene.

regrowth: Natural regeneration of forests as opposed to regeneration with tree planting.

roundwood: Stem wood (logs) as felled in its natural state. It is the base material for sawn wood, plywood, and pulp wood. The "roundwood equivalent" (RWE) is a measure of the volume of logs used in the manufacture of wood products.

RSPO: Roundtable on Sustainable Palm Oil (see box 4.3).

SAR: Synthetic aperture radar (see appendix 2 on satellite sensors).

secondary forest: See box 2.2.

shifting cultivation: Also called swidden agriculture: traditional form of subsistence farming in rotation followed by periods of fallow.

Sierra Club: The oldest and largest American environmental organization.

slash-and-burn farming: Clearing of forest to produce food and cash crops. This is often practiced by immigrant settlers.

smallholder: Small-scale farmer, usually a single family, living from subsistence agriculture and some cash crops.

SPOT: See appendix 2 on satellite sensors.

Stern Review (on the Economics of Climate Change): A report by the economist Nicholas Stern, released by the British government in 2006.

subsistence agriculture: Smallholder farming supporting the needs of a single family.

swidden agriculture: See **shifting cultivation**.

TEEB: "The Economics of Ecosystems and Biodiversity," an international study initiated at the suggestion of the G8+5 countries in Potsdam in 2007, led by Pavan Sukhdev.

teleconnections: A term adopted from climate science to describe factors or events causing an effect in other parts of the world.

terra firme: Rainforest areas in the Amazon Basin that are on firm ground and are not regularly flooded.

Tertiary: A period in the Earth's history between sixty-five million years before present and the beginning of the Quaternary Period 1.8 million years before present.

TFAP: Tropical Forestry Action Plan, created in 1985 (see box 2.1).

TNC: The Nature Conservancy, a large United States–based conservation organization active in over thirty countries.

transect: A path through the forest along which one counts animals or their signs.

TREES: Forest cover monitoring project of the Joint Research Centre of the European Union.

tropical moist forest: See box 1.1.

tropical rainforest: See box 1.1.

UK Met Office Hadley Centre: The United Kingdom's government climate research center.

Unasylva: Forestry journal published by the FAO.

UNCBD: See CBD.

UNCED: The 1992 United Nations Conference on Environment and Development, often referred to as the Rio Conference or the Earth Summit.

UNDP: United Nations Development Programme.

UNEP: United Nations Environment Programme.

UNESCO: United Nations Educational, Scientific and Cultural Organization.

UNFCCC: United Nations Framework Convention on Climate Change.

UNFF: United Nations Forum on Forests (see box 2.1).

várzea: Forest areas that are seasonally flooded by whitewater, nutrient-rich water from the Amazon and Solimões Rivers, in the Amazon Basin.

VPA: Voluntary partnership agreement under the FLEGT action plan.

DESDynI satellite: See appendix 2.

WCMC: World Conservation Monitoring Centre of the UNEP based in Cambridge, United Kingdom.

WCS: Wildlife Conservation Society, founded in 1895 as the New York Zoological Society, based at the Bronx Zoo, active with conservation projects in over sixty countries.

World Commission on Forests and Sustainable Development: See box 2.1.

World Resources Institute (WRI): A global policy and research institute based in Washington, DC, which launched the online forest monitoring system Global Forest Watch in 2014.

World Rainforest Movement: An international nongovernmental organization and network of indigenous groups headquartered in Montevideo, Uruguay (see also box 2.1).

WSSD: World Summit on Sustainable Development, Johannesburg 2002.

WWF: World Wide Fund for Nature (known as the World Wildlife Fund in the United States and Canada) large international nongovernmental organization founded in 1961 and represented in over one hundred countries.

Yaoundé Process: See box 7.2.

zero net deforestation: A target of an equilibrium between deforestation and reforestation.

NOTES

CHAPTER 1

1   D. H. Meadows, D. L. Meadows, J. Randers, and W. W. Behrens, *The Limits to Growth* (New York: Universe Books, 1972).

2   M. Williams, *Deforesting the Earth: From Prehistory to Global Crisis* (Chicago and London: University of Chicago Press, 2003).

3   J. F. Richards, "Land Transformation," in *The Earth as Transformed by Human Action: Global and Regional Changes in the Biosphere Over the Past 300 Years*, eds. B. L. Turner II, W. C. Clark, R. W. Kates, J. F. Richards, J. T. Mathews, and W. B. Meyer (Cambridge University Press, 1990), 163–78.

4   M. Williams, *Deforesting the Earth: From Prehistory to Global Crisis* (Chicago and London: University of Chicago Press, 2003).

5   C. Martin, *The Rainforests of West Africa: Ecology—Threats—Conservation* (Basel, Boston, Berlin: Birkhäuser, 1991).

6   M. Williams, *Deforesting the Earth: From Prehistory to Global Crisis* (Chicago and London: University of Chicago Press, 2003).

7   K. Klein Goldewijk, 2001. "Estimating Global Land Use Change over the Past 300 Years: The HYDE Database," *Global Biogeochemical Cycles* 15, no. 2 (2001): 417–33.

8   J. F. Richards, "Land Transformation," in *The Earth as Transformed by Human Action: Global and Regional Changes in the Biosphere Over the Past 300 Years*, eds. B. L. Turner II, W. C. Clark, R. W. Kates, J. F. Richards, J. T. Mathews, and W. B. Meyer (Cambridge University Press, 1990), 163–78.

9   F. Nectoux and Y. Kuroda, *Timber from the South Seas: An Analysis of Japan's Tropical Timber Trade and Its Environmental Impact* (Gland, Switzerland: WWF International, 1989).

10   E. J. H. Corner, "Suggestions for Botanical Progress," *New Phytologist* 45 (1946): 185–92.

11   P. W. Richards, *The Tropical Rain Forest* (London: Cambridge University Press, 1952).

12   A. Gómez-Pompa, C. Vasquez-Yanes, and C. Guevara, "The Tropical Rain Forest: A Nonrenewable Resource," *Science* 117 (1972): 762–65.

13   W. M. Denevan, "Development and the Imminent Demise of the Amazon Rain Forest," *Professional Geographer* 25 (1973): 130–35.

14   W. M. Denevan, "The Pristine Myth: The Landscape of the Americas in 1492," *Annals of the Association of American Geographers* 82/3 (1992): 369–85.

15   R. Persson, "World Forest Resources: Review of the World's Forest Resources in the early 1970s," in *Department of Forest Survey Research Note No. 17* (Stockholm: Royal College of Forestry, 1974).

16   A. Sommer, "Attempt at an Assessment of the World's Tropical Moist Forests," *Unasylva* 28, no. 112–13 (1976): 5–25.

17   J. B. Hall and M. D. Swaine, "Classification and Ecology of Closed Canopy Forest in Ghana," *Journal of Ecology* 64 (1976): 913–51.

18   C. Martin, *The Rainforests of West Africa: Ecology—Threats—Conservation* (Basel, Boston, Berlin: Birkhäuser, 1991).

19   A. F. W. Schimper, *Pflanzengeographie auf Physiologischer Grundlage* (Jena: G. Fischer, 1898).

20   P. W. Richards, *The Tropical Rain Forest* (London: Cambridge University Press, 1952).

21   T. C. Whitmore, *An Introduction to Tropical Rainforests* (Oxford: Oxford University Press, 1990).

22   UNESCO, *International Classification and Mapping of Vegetation* (Paris: UNESCO, 1973). The UNESCO classification was not just based on climatic factors (climatic climax), but on maps of physiognomic and structural vegetation criteria as well.

23   H. G. Lund, "Definitions of Forest, Deforestation, Afforestation, and Reforestation," http://home.comcast.net/~gyde/DEFpaper.htm.

24   Food and Agriculture Organization of the United Nations, *Second Expert Meeting on Harmonizing Forest-Related Definitions for Use by Various Stakeholders* (Rome: FAO, 2002); Food and Agriculture Organization of the United Nations, *Third Expert Meeting on Harmonizing Forest-Related Definitions for Use by Various Stakeholders* (Rome: FAO, 2005).

25   J. P. Lanly, *Tropical Forest Resources*, FAO Forestry Paper 30 (Rome: FAO, 1982).

26 N. Myers, *Conversion of Tropical Moist Forests: A Report Prepared for the Committee on Research Priorities in Tropical Biology of the National Research Council* (Washington, DC: National Academy of Sciences, 1980).

27 N. Myers, *The Sinking Ark: A New Look at the Problem of Disappearing Species* (Oxford: Pergamon Press, 1979); N. Myers, *A Wealth of Wild Species* (Boulder, Colorado: Westview Press, 1983).

28 UNESCO, *International Classification and Mapping of Vegetation* (Paris: UNESCO, 1973).

29 F. Achard, H. D. Eva, H.-J. Stibig, P. Mayaux, J. Gallego, T. Richards, and J.-P. Malingreau, "Determination of Deforestation Rates of the World's Humid Tropical Forests," *Science* 297 (2002): 999–1002.

30 J. F. Richards, "Land Transformation," in *The Earth as Transformed by Human Action: Global and Regional Changes in the Biosphere Over the Past 300 Years,* eds. B. L. Turner II, W. C. Clark, R. W. Kates, J. F. Richards, J. T. Mathews, and W. B. Meyer (Cambridge University Press, 1990), 163–78.

CHAPTER 2

1 World Resources Institute, "Tropical Forests: A Call for Action," Report of an International Task Force Convened by the World Resources Institute (WRI), the World Bank, and the United Nations Development Programme (UNDP) (Washington, DC: WRI, 1985).

2 M. Colchester and L. Lohman, *The Tropical Forestry Action Plan: What Progress?* (Penang, Malaysia: The World Rainforest Movement, 1990).

3 Food and Agriculture Organization of the United Nations, *State of the World's Forests* 2012 (Rome: FAO, 2012).

4 A. Marcoux, 2000. *Population and Deforestation,* Population and the Environment: A Review and Concepts for Population Programmes Part III (Rome: FAO, 2000).

5 T. Rudel and J. Roper, "The Paths to Rainforest Destruction: Cross-National Patterns of Tropical Deforestation, 1975–1990," *World Development* 25 (1997): 53–65.

6 A. Grainger, "Uncertainty of the Construction of Global Knowledge of Tropical Forests," *Progress in Physical Geography* 34, no. 6 (2010): 811–844; A. Grainger, "The Bigger Picture—Tropical Forest Change in Context, Concept and Practice," in *Reforesting Landscapes: Linking Pattern and Process,* Landscape Series 10, eds. H. Nagendra and J. Southworth (Springer Science and Business Media, 2010).

7 FAO, *Global Forest Resources Assessment 2000: Main Report,* FAO Forestry Paper 140 (Rome: FAO, 2001).

8 FAO, *Global Forest Resources Assessment 2010: Main Report,* FAO Forestry Paper 163 (Rome: FAO, 2010).

9   U. Chokkalingam and W. de Jong, "Secondary Forest: A Working Definition and Typology," *International Forestry Review* 3, no. 1 (2001): 19–26.

10   F. Achard, H. J. Stibig, H. Eva, and P. Mayaux, "Tropical Forest Cover Monitoring in the Humid Tropics—TREES Project," *Tropical Ecology* 43, no. 1 (2002): 9–20.

11   F. Achard, H. Eva, H. J. Stibig, P. Mayaux, J. Gallego, T. Richards, and J.-P. Malingreau, "Determination of Deforestation Rates of the World's Humid Tropical Forests," *Science* 297 (2002): 999–1002.

12   M. Collins, J. A. Sayer, and T. C. Whitmore, *The Conservation Atlas of Tropical Forests: Asia and the Pacific* (London: Macmillan Press, 1991); J. A. Sayer, C. S. Harcourt, and M. Collins, *The Conservation Atlas of Tropical Forests: Africa* (London: Macmillan Press, 1992); C. S. Harcourt and J. A. Sayer, *The Conservation Atlas of Tropical Forests: The Americas* (New York: Simon and Schuster, 1996).

13   F. Achard, H. D. Eva, H.-J. Stibig, P. Mayaux, J. Gallego, T. Richards, and J.-P. Malingreau, "Determination of Deforestation Rates of the World's Humid Tropical Forests," *Science* 297 (2002): 999–1002.

14   F. Achard, H. J. Stibig, H. Eva, and P. Mayaux, "Tropical Forest Cover Monitoring in the Humid Tropics—TREES Project," *Tropical Ecology* 43, no. 1 (2002): 9–20.

15   M. Fagan and R. DeFries, *Measurement and Monitoring of the World's Forests: A Review and Summary of Remote Sensing Technical Capability, 2009–2015* (Washington, DC: Resources for the Future, 2009).

16   P. E. Waggoner, *Forest Inventories: Discrepancies and Uncertainties*, Discussion Paper 09–29 (Washington, DC: Resources for the Future, 2009).

17   P. Mayaux, P. Holmgren, F. Achard, H. Eva, H.-J. Stibig, and A. Branthomme, "Tropical Forest Cover Change in the 1990s and Options for Future Monitoring," *Philosophical Transactions of the Royal Society B* 360 (2005): 373–84.

18   F. Achard and M. C. Hansen, "Use of Earth Observation Technology to Monitor Forests across the Globe," in *Global Forest Monitoring from Earth Observation*, eds. F. Achard and M. C. Hansen (Boca Raton, Florida: CRC Press, 2012), 39–54.

19   D. O. Fuller, "Tropical Forest Monitoring and Remote Sensing: A New Era of Transparency in Forest Governance?," *Singapore Journal of Tropical Geography* 27 (2006): 15–29; M. Fagan and R. DeFries, *Measurement and Monitoring of the World's Forests: A Review and Summary of Remote Sensing Technical Capability, 2009–2015* (Washington, DC: Resources for the Future, 2009).

20   M. Broich, M. C. Hansen, P. Potapov, B. Adusei, E. Lindquist, and S. V. Stehman, "Time-Series Analysis of Multi-Resolution Optical Imagery for Quantifying Forest Cover Loss in Sumatra and Kalimantan, Indonesia," *International Journal of Applied Earth Observation and Geoinformation* 13 (2011): 277–91.

21   M. C. Hansen, P. Potapov, and S. Turubanova, "Use of Coarse-Resolution Imag-
     ery to Identify Hot Spots of Forest Loss at the Global Scale," in *Global Forest
     Monitoring from Earth Observation*, eds. F. Achard and M. C. Hansen (Boca Raton,
     Florida: CRC Press, 2012), 93–109.

22   Y. E. Shimabukuro, J. R. dos Santos, A. R. Formaggio, V. Duarte, and B. F. T.
     Rudorff, "The Brazilian Amazon Monitoring Program: PRODES and DETER Proj-
     ects," in *Global Forest Monitoring from Earth Observation*, eds. F. Achard and M. C.
     Hansen (Boca Raton, Florida: CRC Press, 2012).

23   A. Coca-Castro, L. Reymondin, J. J. Tello, and P. Paz, "La deforestación en la
     región del Amazonas en 8 países según datos del sistema Terra-i," 2013,
     http://www.amazonia-andina.org/sites/default/files/uploads/2013/07/
     informedeforestacion8paises.pdf.

24   M. Flores, U. Lopes da Silva Jr., H. Malone, M. Panuncio, J. C. Riveros, S.
     Rodrigues, R. Silva, S. Valenzuela, D. Arancibia, P. Bara-Neto, and M. Symington,
     WWF's *Living Amazon Initiative: A Comprehensive Approach to Conserving the Largest
     Rainforest and River System on Earth*, WWF Strategy Summary (WWF, 2010), 12.

25   Terra-i, Raw data on Amazon countries' deforestation. Accessed January 2014,
     www.terra-i.org.

26   Bolivia showed particularly worrying deforestation rates. Over the 2004–2012
     period, the deforestation rate in Bolivia was 2.11% of its Amazon biome, exceed-
     ing the deforestation rates of Colombia and Peru almost threefold.

27   M. Fagan and R. DeFries, *Measurement and Monitoring of the World's Forests: A Review
     and Summary of Remote Sensing Technical Capability, 2009–2015* (Washington, DC:
     Resources for the Future, 2009).

28   F. Achard and M. C. Hansen, eds., *Global Forest Monitoring from Earth Observation*
     (Boca Raton, Florida: CRC Press, 2012).

29   M. C. Hansen, P. V. Potapov, R. Moore, M. Hancher, S. A. Turubanova, A. Tyu-
     kavina, D. Thau, S. V. Stehman, S. J. Goetz, T. R. Loveland, A. Kommareddy, A.
     Egorov, L. Chini, C. O. Justice, and J. R. G. Townshend, "High-Resolution Global
     Maps of 21st-Century Forest Cover Change," *Science* 342 (2013): 850–853,
     doi:10.1126/science.1244693.

CHAPTER 3

1   F. Achard and M. C. Hansen, "Use of Earth Observation Technology to Monitor
     Forests across the Globe," in *Global Forest Monitoring from Earth Observation*, eds. F.
     Achard and M. C. Hansen (Boca Raton, Florida: CRC Press, 2012), 39–54.

2   M. C. Hansen, S. Stehman, P. Potapov, T. Loveland, J. Townshend, R. DeFries,
     K. Pittman, B. Arunarwati, F. Stolle, M. Steininger, M. Carroll, and C. DiMiceli,

"Humid Tropical Forest Clearing from 2000 to 2005 Quantified by Using Multi-temporal and Multiresolution Remotely Sensed Data," *Proceedings of the National Academy of Sciences* 105 (2008): 9439–44.

3   C. J. Tucker and J. R. G. Townshend, "Strategies for Monitoring Tropical Defor-estation Using Satellite Data," *International Journal of Remote Sensing* 21, no. 6/7 (2000): 1461–71.

4   M. C. Hansen, P. V. Potapov, R. Moore, M. Hancher, S. A. Turubanova, A. Tyu-kavina, D. Thau, S. V. Stehman, S. J. Goetz, T. R. Loveland, A. Kommareddy, A. Egorov, L. Chini, C. O. Justice, and J. R. G. Townshend, "High-Resolution Global Maps of 21st-Century Forest Cover Change," *Science* 342 (2013): 850–853, doi:10.1126/science.1244693.

5   F. Achard, H. D. Eva, H.-J. Stibig, P. Mayaux, J. Gallego, T. Richards, and J.-P. Mal-ingreau, "Determination of Deforestation Rates of the World's Humid Tropical Forests," *Science* 297 (2002): 999–1002.

6   M. C. Hansen, P. V. Potapov, R. Moore, M. Hancher, S. A. Turubanova, A. Tyu-kavina, D. Thau, S. V. Stehman, S. J. Goetz, T. R. Loveland, A. Kommareddy, A. Egorov, L. Chini, C. O. Justice, and J. R. G. Townshend, "High-Resolution Global Maps of 21st-Century Forest Cover Change," *Science* 342 (2013): 850–853, doi:10.1126/science.1244693.

7   FAO, *Global Forest Resources Assessment 2010: Main Report*, FAO Forestry Paper 163 (Rome: FAO, 2010).

8   F. Achard, H. D. Eva, H.-J. Stibig, P. Mayaux, J. Gallego, T. Richards, and J.-P. Mal-ingreau, "Determination of Deforestation Rates of the World's Humid Tropical Forests," *Science* 297 (2002): 999–1002.

9   M. C. Hansen, P. V. Potapov, R. Moore, M. Hancher, S. A. Turubanova, A. Tyu-kavina, D. Thau, S. V. Stehman, S. J. Goetz, T. R. Loveland, A. Kommareddy, A. Egorov, L. Chini, C. O. Justice, and J. R. G. Townshend, "High-Resolution Global Maps of 21st-Century Forest Cover Change," *Science* 342 (2013): 850–853, doi:10.1126/science.1244693.

10   FAO, *Global Forest Resources Assessment 2010: Main Report*, FAO Forestry Paper 163 (Rome: FAO, 2010).

11   M. C. Hansen, P. V. Potapov, R. Moore, M. Hancher, S. A. Turubanova, A. Tyu-kavina, D. Thau, S. V. Stehman, S. J. Goetz, T. R. Loveland, A. Kommareddy, A. Egorov, L. Chini, C. O. Justice, and J. R. G. Townshend, "High-Resolution Global Maps of 21st-Century Forest Cover Change," *Science* 342 (2013): 850–853, doi:10.1126/science.1244693.

12   M. Broich, M. C. Hansen, P. Potapov, B. Adusei, E. Lindquist, and S. V. Stehman, "Time-Series Analysis of Multi-Resolution Optical Imagery for Quantifying

Forest Cover Loss in Sumatra and Kalimantan, Indonesia," *International Journal of Applied Earth Observation and Geoinformation* 13 (2011): 277–91.

13 M. Broich, M. C. Hansen, F. Stolle, P. Potapov, B. A. Margono, and B. Adusei, "Remotely Sensed Forest Cover Loss Shows High Spatial and Temporal Variation across Sumatra and Kalimantan, Indonesia, 2000–2008," *Environmental Research Letters* 6 (2011): 014010.

14 B. A. Margono, P. V. Potapov, S. Turubanova, F. Stolle, and M. C. Hansen, "Primary Forest Cover Loss in Indonesia over 2000–2012," *Nature Climate Change* 4 (2014): 730–35, doi:10.1038/nclimate2277.

15 M. C. Hansen, P. V. Potapov, R. Moore, M. Hancher, S. A. Turubanova, A. Tyukavina, D. Thau, S. V. Stehman, S. J. Goetz, T. R. Loveland, A. Kommareddy, A. Egorov, L. Chini, C. O. Justice, and J. R. G. Townshend, "High-Resolution Global Maps of 21st-Century Forest Cover Change," *Science* 342 (2013): 850–853, doi:10.1126/science.1244693.

16 D. P. Edwards and W. F. Laurance, "Carbon Emissions: Loophole in Forest Plan for Indonesia," *Nature* 477 (2011): 33, doi:10.1038/477033a.

17 D. Skole and C. Tucker, "Tropical Deforestation and Habitat Fragmentation in the Amazon: Satellite Data from 1978 to 1988," *Science* 260 (1993): 1905–09.

18 Y. E. Shimabukuro, J. R. dos Santos, A. R. Formaggio, V. Duarte, and B. F. T. Rudorff, "The Brazilian Amazon Monitoring Program: PRODES and DETER Projects," in *Global Forest Monitoring from Earth Observation*, eds. F. Achard and M. C. Hansen (Boca Raton, Florida: CRC Press, 2012).

19 D. Nepstad, D. McGrath, J. Jimada, and C. Stickler, "Why Is Amazon Deforestation Climbing?," Mongabay, 17 November, 2013, http://news.mongabay.com/2013/1116-nepstad-why-is-deforestation-climbing.html.

20 P. Potapov, S. Turubanova, M. C. Hansen, I. Zhuravleva, A. Yaroshenko, and L. Laestadius, "Monitoring Forest Loss and Degradation at National to Global Scales Using Landsat Data," in *Global Forest Monitoring from Earth Observation*, eds. F. Achard and M. C. Hansen (Boca Raton, Florida: CRC Press, 2012), 129–152.

21 M. Pereira Goncalves, M. Panjer, T. S. Greenberg, and W. B. Magrath, *Justice for Forests: Improving Criminal Justice Efforts to Combat Illegal Logging*, World Bank study (Washington, DC: World Bank, 2012).

22 CIRAD Blog, "Sylvie Gourlet-Fleury: 'Certaines forêts pourraient produire davantage de bois, d'autres doivent être mieux protégées,'" 9 July, 2013, http://www.cirad.fr/actualites/toutes-les-actualites/articles/2013/questions-a/sylvie-gourlet-fleury-coforchange; A. Fayolle, B. Engelbrecht, V. Freycon, F. Mortier, M. Swaine, M. Réjou-Méchain, J.-L. Doucet, N. Fauvet, G. Cornu, S. Gourlet-Fleury,

"Geological Substrates Shape Tree Species and Trait Distributions in African Moist Forests," PLOS ONE 7 (2012): e42381.

23 P. O. Cerutti and L. Tacconi, "Forests, Illegality, and Livelihoods: The Case of Cameroon," *Society and Natural Resources* 21, no. 9 (2008): 845–853, cited in R. E. Atyi, G. Lescuyer, J. N. Poufoun, and T. M. Fouda, eds., *Étude de l'importance économique et sociale du secteur forestier et faunique au Cameroun* (Yaoundé, Cameroon: CIFOR, 2013).

24 WWF, "Cameroon sends army to defend borders from Sudanese poachers," 16 November, 2012, http://wwf.panda.org/wwf_news/?206742/ Cameroon-sends-army-to-defend-borders-from-Sudanese-poachers.

25 E. Lammerts van Bueren, R. Zagt, and H. Savenije, "Stimulating the Demand for Sustainably Sourced and Licensed Tropical Timber on the European Market," Discussion paper, EU sustainable tropical timber coalition, November 2013.

26 R. Matondo, "La stratégie nationale d'afforestation et de reboisement comme opportunité d'affaire en République du Congo," Présentation au Forum International sur le Développement durable de la Filière Bois dans les Pays du Bassin du Congo, 21–22 October, 2013; C. Luttrell, K. Obidzinski, M. Brockhaus, E. Muharrom, E. Petkova, A. Wardell, and J. Halperin, "Lessons for REDD+ from Measures to Control Illegal Logging in Indonesia," CIFOR Working Paper 74 (Center for International Forestry Research, 2011).

27 Boubacar Bensalah (timber industry association in Abidjan, Côte d'Ivoire), personal communication, 2013; Grid Arendal, "Land Use Change Is Resulting in Rapid Loss of Forests and Excessive Fragmentation: Deforestation in Cote d'Ivoire Case Study," http://www.grida.no/publications/vg/africa/page/3108. aspx.

28 Gavin Neath (Unilever), e-mail exchange with author, 2014.

29 Prosper Obame (Ministry of Forestry, Libreville, Gabon), e-mail exchange with author, 2014.

30 Alain Ngoya-Kessy (Ministry of Forest Economy, Republic of the Congo), e-mail exchange with author, 2014.

31 Jean-Avit Kongape (Ministry of Forestry, Cameroon), e-mail exchange with author, 2014.

32 James Acworth (World Bank office in Yaoundé, Cameroon), personal communication, 2012.

33 *Forests News*, "'Landscape Approaches' Can End the Debate That Pits Agriculture against Forests, Say Experts," blog entry by C. Moss, 22 June, 2012,

http://blog.cifor.org/9829/landscape-approaches-can-end-the-debate-that-pits-agriculture-against-forests-say-experts#.VAAr043lr5w.

34   S. Brown and D. Zarin, "What Does Zero Deforestation Mean?," *Science* 342 (2013): 805–7.

SPECIALIST'S VIEW: SAYER

1   J. Sayer, J. Ghazoul, P. Nelson, and A. K. Boedhihartono, "Oil Palm Expansion Transforms Tropical Landscapes and Livelihoods," *Global Food Security* 1, no. 2 (2012): 114–19, doi:10.1016/j.gfs.2012.10.003.

CHAPTER 4

1   D. Nepstad, C. M. Stickler, and O. T. Almeida, "Globalization of the Amazon Soy and Beef Industries: Opportunities for Conservation," *Conservation Biology* 20, no. 6 (2006): 1595–1603.

2   H. J. Geist and E. F. Lambin, "Proximate Causes and Underlying Driving Forces of Tropical Deforestation, *BioScience* 52, no. 2 (2002): 143–50.

3   H. K. Gibbs, A. S. Ruesch, F. Achard, M. K. Clayton, P. Holmgren, N. Ramankutty, and J. A. Foley, "Tropical Forests Were the Primary Sources of New Agricultural Land in the 1980s and 1990s," *Proceedings of the National Academy of Sciences* 107, no. 38 (2010): 16732–37.

4   J. Blaser and C. Robledo, "Initial Analysis on the Mitigation Potential in the Forestry Sector," Report prepared for the UNFCCC Secretariat, 2007, http://unfccc.int/files/cooperation_and_support/financial_mechanism/application/pdf/blaser.pdf; N. Hosonuma, M. Herold, V. DeSy, R. S. DeFries, M. Brockhaus, L. Verchot, A. Angelsen, and E. Romijn, "An Assessment of Deforestation and Forest Degradation Drivers in Developing Countries," *Environmental Research Letters* 7 (2012): 044009, doi:10.1088/1748–9326/7/4/044009.

5   N. Hosonuma, M. Herold, V. DeSy, R. S. DeFries, M. Brockhaus, L. Verchot, A. Angelsen, and E. Romijn, "An Assessment of Deforestation and Forest Degradation Drivers in Developing Countries," *Environmental Research Letters* 7 (2012): 044009, doi:10.1088/1748-9326/7/4/044009.

6   T. Rudel, "The National Determinants of Deforestation in Sub-Saharan Africa," *Philosophical Transactions of the Royal Society B* 368 (2013): 20120405, doi:10.1098/rstb.2012.0405.

7   P. Potapov, S. Turubanova, M. C. Hansen, I. Zhuravleva, A. Yaroshenko, and L. Laestadius, "Monitoring Forest Loss and Degradation at National to Global

Scales Using Landsat Data," in *Global Forest Monitoring from Earth Observation*, eds. F. Achard and M. C. Hansen (Boca Raton, Florida: CRC Press, 2012), 129–152.

8   J. Schure, V. Ingram, J.-N. Marien, R. Nasi, and E. Dubiez, *Woodfuel for Urban Centres in the Democratic Republic of Congo*, CIFOR Brief No. 7 (Center for International Forestry Research, November 2011).

9   P. Mayaux, J.-F. Pekel, B. Desclée, F. Donnay, A. Lupi, F. Achard, M. Clerici, C. Bodart, A. Brink, R. Nasi, and A. Belward, "State and Evolution of the African Rainforests between 1990 and 2010," *Philosophical Transactions of the Royal Society B* 368 (2013): 20120300, doi:10-1098/rstb.2012.0300.

10 P. Sist, N. Picard, and S. Gourlet-Fleury, "Sustainable Cutting Cycle and Yields in a Lowland Mixed Dipterocarp Forest of Borneo," *Annals of Forest Science* 60 (2003): 803–814, doi:10.1051/forest:2003075.

11 H. J. Geist and E. F. Lambin, "Proximate Causes and Underlying Driving Forces of Tropical Deforestation, *BioScience* 52, no. 2 (2002): 143–50.

12 R. A. Butler, "Rainforest Logging," Mongabay, last updated 27 July, 2012, http://rainforests.mongabay.com/0807.htm.

13 W. Laurance, S. Bergen, M. Cochrane, P. Fearnside, P. Delamônica, S. Agra d'Angelo, C. Barber, and T. Fernandes, "The Future of the Amazon," in *Tropical Rainforests–Past, Present, and Future*, eds. E. Bermingham, C. Dick, and C. Moritz (Chicago: University of Chicago Press, 2005), 583–609.

14 N. Bayol, B. Demarquez, C. de Wasseige, R. Eba'a Atyi, J.-F. Fisher, R. Nasi, A. Pasquier, X. Rossi, M. Steil, and C. Vivien, "Forest Management and the Timber Sector in Central Africa," in *The Forests of the Congo Basin—State of the Forest* 2010, eds. C. de Wasseige, P. de Marcken, N. Bayol, F. Hiol Hiol, P. Mayaux, B. Desclée, R. Nasi, A. Billand, P. Defourny, and R. Eba'a Atyi (Luxembourg: Publications Office of the European Union, 2012), 43–62.

15 P. Mayaux, J.-F. Pekel, B. Desclée, F. Donnay, A. Lupi, F. Achard, M. Clerici, C. Bodart, A. Brink, R. Nasi, and A. Belward, "State and Evolution of the African Rainforests between 1990 and 2010," *Philosophical Transactions of the Royal Society B* 368 (2013): 20120300, doi:10-1098/rstb.2012.0300.

16 COMIFAC (Commission des Forêts d'Afrique Centrale), the Central African Forest Commission, was legally created by the Yaoundé Declaration of the Central African heads of state meeting in March 1999, and formally established in 2005 as a regional forum for the conservation and sustainable management of the Congo Basin forests.

17 Y. Malhi, S. Adu-Bredu, R. A. Asare, S. L. Lewis, and P. Mayaux, "African Rainforests: Past, Present and Future," *Philosophical Transactions of the Royal Society B* 368 (2013): 20120312.

18 C. de Wasseige, P. de Marcken, N. Bayol, F. Hiol Hiol, P. Mayaux, B. Desclée, R. Nasi, A. Billand, P. Defourny, and R. Eba'a Atyi, eds., *The Forests of the Congo Basin— State of the Forest 2010* (Luxembourg: Publications Office of the European Union, 2012).

19 Y. Malhi, S. Adu-Bredu, R. A. Asare, S. L. Lewis, and P. Mayaux, "African Rainforests: Past, Present and Future," *Philosophical Transactions of the Royal Society B 368* (2013): 20120312; P. Mayaux, J.-F. Pekel, B. Desclée, F. Donnay, A. Lupi, F. Achard, M. Clerici, C. Bodart, A. Brink, R. Nasi, and A. Belward, "State and Evolution of the African Rainforests between 1990 and 2010," *Philosophical Transactions of the Royal Society B 368* (2013): 20120300, doi:10-1098/rstb.2012.0300.

20 G. P. Asner, T. Rudel, T. Mitchell Aide, R. DeFries, and R. Emerson, "A Contemporary Assessment of Change in Humid Tropical Forests," *Conservation Biology 23*, no. 6 (2009): 1386–95.

21 C. Martin, *The Rainforests of West Africa: Ecology—Threats—Conservation* (Basel, Boston, Berlin: Birkhäuser, 1991).

22 C. J. Clark, J. R. Poulsen, R. Malonga, and P. W. Elkan, "Logging Concessions Can Extend the Conservation Estate for Central African Tropical Forests," *Conservation Biology 23*, no. 5 (2009): 1281–93, doi:10.1111/j.1523–1739.2009.01243.x.

23 The definition of "illegal logging" varies from narrow to very broad. It can refer to irregularities in harvesting, transport, processing, and trade through the entire chain of custody, and includes evasion of taxes and fees. A World Bank study distinguished three aspects concerning illegal harvesting alone: 1) illegal products: felling of protected tree species, felling undersized trees, etc.; 2) illegal location: felling in areas where logging is prohibited, or without a valid permit, or in protected sites within logging concessions; and 3) illegal practices such as failing to file forest management plans, failing to carry out impact assessments, or failing to carry out post-harvest reforestation.

24 S. Lawson and L. Macfaul, *Illegal Logging and Related Trade: Indicators of the Global Response* (London: Chatham House, 2010).

25 The World Bank, *Strengthening Law Enforcement and Governance: Addressing a Systemic Constraint to Sustainable Development*, Report No. 36638-GLB (Washington, DC: World Bank, 2006).

26 M. Pereira Goncalves, M. Panjer, T. S. Greenberg, and W. B. Magrath, *Justice for Forests: Improving Criminal Justice Efforts to Combat Illegal Logging*, World Bank study (Washington, DC: World Bank, 2012).

27 L. Neme, "Top Officials Busted in Amazon Logging Raids, but Political Patronage May Set Them Free," Mongabay, 8 July, 2010, http://news. mongabay. com/2010/0708-neme_operation_jurupari.html.

28 S. Lawson and L. Macfaul, *Illegal Logging and Related Trade: Indicators of the Global Response* (London: Chatham House, 2010).

29 S. Lawson and L. Macfaul, *Illegal Logging and Related Trade: Indicators of the Global Response* (London: Chatham House, 2010).

30 Mongabay, "Gibson Guitar to Pay $300,000 for Violating Lacey Act with Illegal Timber Imports from Madagascar," 6 August, 2012, http://news.mongabay.com/2012/0806-gibson-doj-lacey.html.

31 Chatham House maintains the Illegal Logging Portal, a website that provides an overview of key issues and includes a searchable database, at http://www.illegal-logging.info.

32 S. Lawson and L. Macfaul, *Illegal Logging and Related Trade: Indicators of the Global Response* (London: Chatham House, 2010).

33 C. Martin, *The Rainforests of West Africa: Ecology—Threats—Conservation* (Basel, Boston, Berlin: Birkhäuser, 1991).

34 wwf, "About Pulp & Paper Production and Use," http://wwf.panda.org/what_we_do/footprint/forestry/sustainablepulppaper/aboutpulppaperproductionuse.

35 Eyes on the Forest, "The Truth behind app's Greenwash," 14 December, 2011, http://www.eyesontheforest.or.id/attach/EoF%20(14Dec11)%20The%20truth%20behind%20apps%20greenwash%20hr.pdf.

36 R. A. Butler, "Asia Pulp & Paper Hires Top U.S. Lobbyist to Help 'Green' Its Image," Mongabay, 5 December, 2012, http://news.mongabay.com/2012/1205-app-lobbying-eizenstat.html#5lkBQCBJ4FdsvbqB.99.

37 wwf, "Can Indonesia's Notorious Deforesters Turn Over a New Leaf?," 6 February, 2014, http://worldwildlife.org/stories/can-indonesia-s-notorious-deforesters-turn-over-a-new-leaf.

38 F. Mousseau, *On Our Land: Modern Land Grabs Reversing Independence in Papua New Guinea* (Oakland, California: Oakland Institute, 2013).

39 N. Myers, *Tropical Moist Forests: Over-Exploited and Under-Utilized?*, Report to wwf/iucn (Gland, Switzerland: wwf International, 1985).

40 C. Martin, *The Rainforests of West Africa: Ecology—Threats—Conservation* (Basel, Boston, Berlin: Birkhäuser, 1991).

41 aocs Lipid Library, "Oils and Fats in the Market Place: Commodity Oils and Fats: Palm Oil," http://lipidlibrary.aocs.org/market/palmoil.htm.

42 E. Saxon and S. Roquemore, "Palm Oil," in *The Root of the Problem–What's Driving Tropical Deforestation Today?*, eds. D. Boucher, P. Elias, K. Lininger, C. May-Tobin,

S. Roquemore, and E. Saxon (Cambridge, Massachusetts: Union of Concerned Scientists, 2011), http://www.ucsusa.org/global_warming/solutions/stop-defor-estation/drivers-of-deforestation.html.

43   International Finance Corporation, *The World Bank Group Framework and IFC Strategy for Engagement in the Palm Oil Sector* (World Bank, 2011), http://www.ifc.org/wps/wcm/connect/159dce004ea3bd0fb359f71dc0e8434d/WBG+Frame-work+and+IFC+Strategy_FINAL_FOR+WEB.pdf?MOD=AJPERES.

44   E. Saxon and S. Roquemore, "Palm Oil," in *The Root of the Problem–What's Driving Tropical Deforestation Today?*, eds. D. Boucher, P. Elias, K. Lininger, C. May-Tobin, S. Roquemore, and E. Saxon (Cambridge, Massachusetts: Union of Concerned Scientists, 2011), http://www.ucsusa.org/global_warming/solutions/stop-defor-estation/drivers-of-deforestation.html.

45   D. Sheil, A. Casson, E. Meijaard, M. van Noordwijk, J. Gaskell, J. Sunderland-Groves, K. Wertz, and M. Kanninen, "The Impacts and Opportunities of Oil Palm in Southeast Asia: What Do We Know and What Do We Need to Know?," Occasional paper (Bogor, Indonesia: Center for International Forestry Research, 2009).

46   I. Gerasimchuk and P. Y. Koh, "The EU Biofuel Policy and Palm Oil: Cutting Subsidies or Cutting Rainforest?," Research report (Winnipeg, Canada: Global Subsidies Initiative, International Institute for Sustainable Development, 2013).

47   J. Sayer, J. Ghazoul, P. Nelson, and A. K. Boedhihartono, "Oil Palm Expansion Transforms Tropical Landscapes and Livelihoods," *Global Food Security* 1, no. 2 (2012): 114–119, doi:10.1016/j.gfs.2012.10.003.

48   FAOSTAT, http://faostat3.fao.org/faostat-gateway/go/to/download/Q/QC/E.

49   J. Sayer, J. Ghazoul, P. Nelson, and A. K. Boedhihartono, "Oil Palm Expansion Transforms Tropical Landscapes and Livelihoods," *Global Food Security* 1, no. 2 (2012): 114–119, doi:10.1016/j.gfs.2012.10.003.

50   E. Saxon and S. Roquemore, "Palm Oil," in *The Root of the Problem–What's Driving Tropical Deforestation Today?*, eds. D. Boucher, P. Elias, K. Lininger, C. May-Tobin, S. Roquemore, and E. Saxon (Cambridge, Massachusetts: Union of Concerned Scientists, 2011), http://www.ucsusa.org/global_warming/solutions/stop-defor-estation/drivers-of-deforestation.html.

51   D. Sheil, A. Casson, E. Meijaard, M. van Noordwijk, J. Gaskell, J. Sunderland-Groves, K. Wertz, and M. Kanninen, "The Impacts and Opportunities of Oil Palm in Southeast Asia: What Do We Know and What Do We Need to Know?," Occa-sional paper (Bogor, Indonesia: Center for International Forestry Research, 2009).

52 I. Gerasimchuk and P. Y. Koh, "The EU Biofuel Policy and Palm Oil: Cutting Subsidies or Cutting Rainforest?," Research report (Winnipeg, Canada: Global Subsidies Initiative, International Institute for Sustainable Development, 2013).

53 R. Carrere, Oil Palm in Africa–Past, Present and Future Scenarios, WRM Series No. 15 (World Rainforest Movement, 2010).

54 L. Feintrenie, S. Schwarze, and P. Levang, "Are Local People Conservationists? Analysis of Transition Dynamics from Agroforests to Monoculture Plantations in Indonesia," Ecology and Society 15, no. 4 (2010): article 37.

55 J. Sayer, J. Ghazoul, P. Nelson, and A. K. Boedhihartono, "Oil Palm Expansion Transforms Tropical Landscapes and Livelihoods," Global Food Security 1, no. 2 (2012): 114–19, doi:10.1016/j.gfs.2012.10.003.

56 D. Boucher, "Soybeans," in The Root of the Problem–What's Driving Tropical Deforestation Today?, eds. D. Boucher, P. Elias, K. Lininger, C. May-Tobin, S. Roquemore, and E. Saxon (Cambridge, Massachusetts: Union of Concerned Scientists, 2011), http://www.ucsusa.org/global_warming/solutions/stop-deforestation/drivers-of-deforestation.html.

57 R. A. Butler, "The Impact of Industrial Agriculture in Rainforests," Mongabay, 28 July, 2012, http://rainforests.mongabay.com/0811.htm.

58 D. Boucher, "Soybeans," in The Root of the Problem–What's Driving Tropical Deforestation Today?, eds. D. Boucher, P. Elias, K. Lininger, C. May-Tobin, S. Roquemore, and E. Saxon (Cambridge, Massachusetts: Union of Concerned Scientists, 2011), http://www.ucsusa.org/global_warming/solutions/stop-deforestation/drivers-of-deforestation.html.

59 D. C. Morton, R. S. DeFries, Y. E. Shimabukuro, L. O. Anderson, E. Arai, F. del Bon Espirito-Santo, R. Freitas, and J. Morisette, "Cropland Expansion Changes Deforestation Dynamics in the Southern Brazilian Amazon," Proceedings of the National Academy of Sciences 103, no. 39 (2006): 14637–41.

60 FBOMS Forests Work Group, Relation between Expansion of Soy Plantations and Deforestation (Brazilian Forum of NGOs and Social Movements for Environment and Development), http://commodityplatform.org/wp/wp-content/uploads/2007/09/relation-between-expansion-of-soy-plantations-and-deforestation.pdf.

61 D. Nepstad, C. M. Stickler, and O. T. Almeida, "Globalization of the Amazon Soy and Beef Industries: Opportunities for Conservation," Conservation Biology 20, no. 6 (2006): 1595–1603.

62 Greenpeace International, Eating Up the Amazon, 6 April, 2006, http://www.greenpeace.org/usa/Global/usa/report/2010/2/eating-up-the-amazon.pdf.

63 R. A. Butler, "The Impact of Industrial Agriculture in Rainforests," Mongabay, 28 July, 2012, http://rainforests.mongabay.com/0811.htm.

64 PRODES, http://www.dpi.inpe.br/prodesdigital/prodes.php.

65 M. N. Macedo, R. S. DeFries, D. C. Morton, C. M. Stickler, G. L. Galford, and Y. E. Shimabukuro, "Decoupling of Deforestation and Soy Production in the Southern Amazon During the Late 2000s," *Proceedings of the National Academy of Sciences* 109, no. 4 (2012): 1341–46.

66 M. N. Macedo, R. S. DeFries, D. C. Morton, C. M. Stickler, G. L. Galford, and Y. E. Shimabukuro, "Decoupling of Deforestation and Soy Production in the Southern Amazon During the Late 2000s," *Proceedings of the National Academy of Sciences* 109, no. 4 (2012): 1341–46.

67 D. Boucher, "Cattle and Pasture," in *The Root of the Problem–What's Driving Tropical Deforestation Today?*, eds. D. Boucher, P. Elias, K. Lininger, C. May-Tobin, S. Roquemore, and E. Saxon (Cambridge, Massachusetts: Union of Concerned Scientists, 2011) http://www.ucsusa.org/global_warming/solutions/stop-deforestation/drivers-of-deforestation.html.

68 S. Wirsenius, F. Hedenus, and K. Mohlin, "Greenhouse Gas Taxes on Animal Food Products: Rationale, Tax Scheme and Climate Mitigation Effects," *Climatic Change* 108, no. 1–2 (2011): 159-184, doi:10.1007/s10584-010-9971-x.

69 D. Kaimowitz, B. Mertens, S. Wunder, and P. Pacheco, *Hamburger Connection Fuels Amazon Destruction–Cattle Ranching and Deforestation in Brazil's Amazon* (Bogor, Indonesia: Center for International Forestry Research, 2004), http://www.cifor.org/publications/pdf_files/media/Amazon.pdf.

70 S. Wirsenius, F. Hedenus, and K. Mohlin, "Greenhouse Gas Taxes on Animal Food Products: Rationale, Tax Scheme and Climate Mitigation Effects," *Climatic Change* 108, no. 1–2 (2011): 159–84, doi:10.1007/s10584-010-9971-x.

71 D. Kaimowitz, B. Mertens, S. Wunder, and P. Pacheco, *Hamburger Connection Fuels Amazon Destruction–Cattle Ranching and Deforestation in Brazil's Amazon* (Bogor, Indonesia: Center for International Forestry Research, 2004), http://www.cifor.org/publications/pdf_files/media/Amazon.pdf.

72 B. Soares Filho, D. Nepstad, L. Curran, G. Cerqueira, R. Garcia, C. Ramos, E. Voll, A. McDonald, P. Lefebvre, and P. Schlesinger, "Modeling Conservation in the Amazon Basin," *Nature* 440 (2006): 520–23.

73 Greenpeace International, *Slaughtering the Amazon*, 1 June, 2009, http://www.greenpeace.org/international/Global/international/planet-2/report/2009/7/slaughtering-the-amazon.pdf.

74 R. A. Butler, "Brazilian Beef Giant Announces Moratorium on Rainforest Beef," Mongabay, 13 August, 2009, http://news.mongabay.com/2009/0813-bertin_moratorium.html.

75 The Leather Working Group, http://www.leatherworkinggroup.com.

76 Sierra Club, "Minerva Beef Protest Letter," 8 January, 2014, goo.gl/vdctpX.

77 D. Nepstad, D. McGrath, J. Jimada, and C. Stickler, "Why Is Amazon Deforestation Climbing?," Mongabay, 17 November, 2013, http://news.mongabay.com/2013/1116-nepstad-why-is-deforestation-climbing.html.

78 T. Rudel, "Changing Agents of Deforestation: From State-Initiated to Enterprise Driven Processes, 1970–2000," *Land Use Policy* 24 (2007): 35–41.

79 R. S. DeFries, T. Rudel, M. Uriarte, and M. C. Hansen, "Deforestation Driven by Urban Population Growth and Agricultural Trade in the Twenty-First Century," *Nature Geoscience* 3 (2010): 178–81.

80 S. J. Wright and H. C. Muller-Landau, "The Future of Tropical Forest Species," *Biotropica* 38, no. 3 (2006): 287–301.

81 D. Hofstrand, "More on Feeding Nine Billion People by 2050," AgMRC *Renewable Energy and Climate Change Newsletter*, January 2012, http://www.agmrc.org/file.cfm/media/newsletters/AgMRC_012012_20902E1B3E556.pdf.

82 W. F. Laurance, J. Sayer, and K. G. Cassman, "Agricultural Expansion and Its Impacts on Tropical Nature," *Trends in Ecology & Evolution* 29, no. 2 (2013): 107–16.

SPECIALIST'S VIEW: BLASER

1 J. Blaser and I. Thompson, "CPF–Summary Paper on Sustainable Forest Management" (Discussion paper to the attention of the meeting of the Collaborative Partnership on Forests, 28–29 April, 2010).

2 See Specialist's View on page 235 for more about REDD+. With respect to the present discussion, the part of the REDD+ definition that refers to "sustainable management of forests" is of particular interest. It refers to managing existing carbon stocks in natural forests that are subject to timber management.

3 WCFSD, "Our Forests: Our Future," Summary Report of the World Commission on Forests and Sustainable Development, 1999, http://www.iisd.org/pdf/wcfsd-summary.pdf.

4 The ITTO (International Tropical Timber Organization) is an international organization under the United Nations Conference for Trade and Development that promotes forest development and the trade of tropical timber from sustainably managed forests in the tropics.

5   ITTO, *Revised ITTO Criteria and Indicators for the Sustainable Management of Tropical Forests*, ITTO Policy Development Series No 15 (ITTO, 2005).

6   J. Blaser, A. Sarre, D. Poore, and S. Johnson, *Status of Tropical Forest Management 2011*, ITTO Technical Series No. 38 (Yokohama, Japan: ITTO, 2011).

7   J. Blaser and C. Sabogal, ITTO *Guidelines for the Restoration, Management and Rehabilitation of Degraded and Secondary Tropical Forests*, ITTO Policy Development Series No. 13 (ITTO, 2002).

8   J. Blaser, A. Sarre, D. Poore, and S. Johnson, *Status of Tropical Forest Management 2011*, ITTO Technical Series No. 38 (Yokohama, Japan: ITTO, 2011).

9   J. Blaser, A. Sarre, D. Poore, and S. Johnson, *Status of Tropical Forest Management 2011*, ITTO Technical Series No. 38 (Yokohama, Japan: ITTO, 2011).

10   H. Lamprecht, *Waldbau in den Tropen* (Hamburg, Germany: Paul Parey, 1986).

11   F. E. Putz, P. Zuidema, T. Synnott, M. Peña-Claros, M. Pinard, D. Sheil, J. Vanclay, P. Sist, S. Gourlet-Fleury, B. Griscom, J. Palmer, and R. Zagt, "Sustaining Conservation Values in Selectively Logged Tropical Forests: The Attained and the Attainable," *Conservation Letters* 5 (2012): 296–303.

12   R. Nasi and P. Frost, "Sustainable Forest Management in the Tropics: Is Everything in Order but the Patient Still Dying?," *Ecology and Society* 14, no. 2 (2009): 40, http://www.ecologyandsociety.org/vol14/iss2/art40.

13   G. P. Asner, E. N. Broadbent, P. J. Oliveira, M. Keller, D. E. Knapp, and J. N. M. Silva, "Condition and Fate of Logged Forests in the Brazilian Amazon," *Proceedings of the National Academy of Science of the United States of America* 103, no. 34 (2006): 12947–50.

14   L. Curran, S. Trigg, A. McDonald, D. Astiani, Y. M. Hardiono, P. Siregar, I. Caniago, and E. Kasischke, "Lowland Forest Loss in Protected Areas of Indonesian Borneo," *Science* 303 (2004): 1000–03.

15   N. Byron and T. Costantini, "The Economics of Ecologically Sustainable Forest Management and Wildlife Conservation in Tropical Forests," Unpublished manuscript available at CIFOR, Bogor, 1998.

16   S. Higman, et al., *The Sustainable Forestry Handbook* (London: Earthscan Books, 1999).

17   F. E. Putz, P. Zuidema, T. Synnott, M. Peña-Claros, M. Pinard, D. Sheil, J. Vanclay, P. Sist, S. Gourlet-Fleury, B. Griscom, J. Palmer, and R. Zagt, "Sustaining Conservation Values in Selectively Logged Tropical Forests: The Attained and the Attainable," *Conservation Letters* 5 (2012): 296–303.

18   J. Blaser and C. Sabogal, ITTO *Guidelines for the Restoration, Management and Rehabilitation of Degraded and Secondary Tropical Forests*, ITTO Policy Development Series No. 13 (ITTO, 2002).

19 Enrichment planting: planting of desired tree species in managed tropical forests with the objective of creating a high forest dominated by the desirable species.

20 J. Carle and P. Holmgren, "Wood from Planted Forests: A Global Outlook 2005–2030," *Forest Products Journal* 58, no. 12 (2008): 6–18.

21 J. Blaser and H. Gregersen, "Forests in the Next 300 Years," *Unasylva* 64, no. 240 (2013): 61–73.

22 J. Blaser and H. Gregersen, "Forests in the Next 300 Years," *Unasylva* 64, no. 240 (2013): 61–73.

CHAPTER 5

1   A. Mather, "The forest transition," *Area* 24 (1992): 367–79; A. Mather, "The Transition from Deforestation to Reforestation in Europe," in *Agricultural Technologies and Tropical Deforestation*, eds. A. Angelsen and D. Kaimowitz (Wallingford, UK: CABI, 2001), 35–52.

2   A. Mather and J. Fairbairn, "From Floods to Reforestation: The Forest Transition in Switzerland," *Environment and History* 6 (2000): 399–421.

3   Swiss Federal Office for the Environment, "The Swiss Forest in Brief," http://www.bafu.admin.ch/wald/01198/01199/index.html?lang=en.

4   A. Grainger, "Uncertainty in the Construction of Global Knowledge of Tropical Forests," *Progress in Physical Geography* 34, no. 6 (2010): 811–44.

5   A. Angelsen, M. Brockhaus, M. Kanninen, E. Sills, W. D. Sunderlin, and S. Wertz-Kanounnikoff, eds., *Realising REDD+: National Strategy and Policy Options* (Bogor, Indonesia: CIFOR, 2009), 1–9.

6   T. Rudel, O. Coomes, E. Moran, F. Achard, A. Angelsen, J. Xu, and E. Lambin, "Forest Transitions: Towards a Global Understanding of Land Use Change," *Global Environmental Change* 15 (2005): 23–31.

7   C. Zhu, R. Taylor, and G. Feng, *China's Wood Market, Trade and the Environment* (Monmouth Junction, New Jersey: Science Press USA, 2004).

8   S. Hecht and S. Saatchi, "Globalization and Forest Resurgence: Changes in Forest Cover in El Salvador," *Bioscience* 57, no. 8 (2007): 663–72.

9   S. Hecht and S. Saatchi, "Globalization and Forest Resurgence: Changes in Forest Cover in El Salvador," *Bioscience* 57, no. 8 (2007): 663–72.

10 T. Rudel, O. Coomes, E. Moran, F. Achard, A. Angelsen, J. Xu, and E. Lambin, "Forest Transitions: Towards a Global Understanding of Land Use Change," *Global Environmental Change* 15 (2005): 23–31.

11   C. Kull, C. Ibrahim, and T. Meredith, "Tropical Forest Transitions and Global-
     ization: Neoliberalism, Migration, Tourism, and International Conservation
     Agendas," *Society and Natural Resources* 20, no. 8 (2007): 723–37.

12   D. K. Munroe, J. Southworth, and C. M. Tucker, "Modeling Spatially and Tempo-
     rally Complex Land-Cover Change: The Case of Western Honduras," *Professional
     Geographer* 56, no. 4 (2004): 544–59.

13   S. B. Roy, "Forest Protection Committees in West Bengal," *Economic and Political
     Weekly* 27, no. 29 (1992): 1528–30.

14   N. H. Ravindranath, R. K. Chaturvedi, and I. K. Murthy, "Forest Conservation,
     Afforestation and Reforestation in India: Implications for Forest Carbon Stocks,"
     Current Science 95, no. 2 (2008): 216–22.

15   P. Meyfroidt and E. Lambin, "Forest Transition in Vietnam and Its Environmen-
     tal Impacts," *Global Change Biology* 14, no. 6 (2008): 1319–36.

16   G. P. Asner, T. Rudel, T. Mitchell Aide, R. DeFries, and R. Emerson, "A Contempo-
     rary Assessment of Change in Humid Tropical Forests," *Conservation Biology* 23,
     no. 6 (2009): 1386–95.

17   F. Achard, H. D. Eva, H.-J. Stibig, P. Mayaux, J. Gallego, T. Richards, and J.-P. Mal-
     ingreau, "Determination of Deforestation Rates of the World's Humid Tropical
     Forests," *Science* 297 (2002): 999–1002.

18   T. M. Aide, M. Clark, H. Grau, D. López-Carr, M. Levy, D. Redo, M. Bonilla-
     Moheno, G. Riner, M. Andrade-Núñez, and M. Muñiz, "Deforestation and Refor-
     estation of Latin America and the Caribbean (2001–2010)," *Biotropica* 45, no. 2
     (2012): 1–10.

19   K. Ehrhardt-Martinez, E. Crenshaw, and C. Jenkins, "Deforestation and the
     Environmental Kuznets Curve: A Cross-National Investigation of Intervening
     Mechanisms," *Social Science Quarterly* 83, no. 1 (2002): 226–43.

20   J. H. Mills, and T. A. Waite, "Economic Prosperity, Biodiversity Conservation,
     and the Environmental Kuznets Curve," *Ecological Economics* 68, no. 7 (2009):
     2087–95; C. Martin, "Globalization and National Environmental Policy: The
     Influence of WWF, an International Non-governmental Organization," in *A
     Handbook of Globalisation and Environmental Policy*, eds. F. Wijen, K. Zoeteman, and
     J. Pieters (Cheltenham, UK: Edward Elgar, 2006), 371–93.

21   P. Meyfroidt, T. Rudel, and E. Lambin, "Forest Transitions, Trade, and the Global
     Displacement of Land Use," *Proceedings of the National Academy of Sciences* 107,
     no. 49 (2010): 20917–22.

22   T. Rudel, L. Schneider, and M. Uriarte, "Forest Transitions: An Introduction,"
     Land Use Policy 27 (2010): 95–97.

23 C. Kull, C. Ibrahim, and T. Meredith, "Tropical Forest Transitions and Global-
    ization: Neoliberalism, Migration, Tourism, and International Conservation
    Agendas," *Society and Natural Resources* 20, no. 8 (2007): 723–37.

24 P. Meyfroidt, T. Rudel, and E. Lambin, "Forest Transitions, Trade, and the Global
    Displacement of Land Use," *Proceedings of the National Academy of Sciences* 107,
    no. 49 (2010): 20917–22.

25 P. Meyfroidt, M. van Noordwijk, P. A. Minang, S. Dewi, and E. Lambin, "Drivers
    and Consequences of Tropical Forest Transitions: Options to Bypass Land Deg-
    radation?," ASB Policy Brief 25 (Nairobi, Kenya: ASB Partnership for the Tropical
    Forest Margins, 2011).

26 J. Southworth and H. Nagendra, "Reforestation: Challenges and Themes in
    Reforestation Research," in *Reforesting Landscapes: Linking Pattern and Process*,
    Landscape Series 10, eds. H. Nagendra and J. Southworth (Springer Science and
    Business Media, 2010), 1–14.

27 T. Rudel, O. Coomes, E. Moran, F. Achard, A. Angelsen, J. Xu, and E. Lambin, "For-
    est Transitions: Towards a Global Understanding of Land Use Change," *Global
    Environmental Change* 15 (2005): 23–31.

28 S. Brown and D. Zarin, "What Does Zero Deforestation Mean?," *Science* 342
    (2013): 805–7.

CHAPTER 6

1   E. O. Wilson, ed., *Biodiversity*, (Washington, DC: National Academy Press, 1988).

2   IUCN, World Resources Institute, World Conservation Union, and United
    Nations Environment Programme, *Global Biodiversity Strategy–Guidelines for Action
    to Save, Study, and Use Earth's Biotic Wealth Sustainably and Equitably* (Washington, DC:
    World Resources Institute, 1992).

3   Organisation for Economic Co-operation and Development, *Environmen-
    tal Outlook to 2050: The Consequences of Inaction* (OECD Publishing, 2012),
    doi:10.1787/9789264122246-en.

4   R. M. May, "Why Worry about How Many Species and Their Loss?," PLOS
    Biology 9, no. 8 (2011), doi:10.1371/journal.pbio.1001130.

5   T. L. Erwin, "Tropical Forests: Their Richness in Coleoptera and Other Arthro-
    pod Species," *Coleoptera Bulletin* 36 (1982): 74–75; T. L. Erwin, "Beetles and Other
    Insects of Tropical Forest Canopies at Manaus, Brazil, Sampled by Insecticidal
    Fogging," in *Tropical Rainforest: Ecology and Management*, eds. S. L. Sutton, T. C.
    Whitmore, and A. C. Chadwick (Oxford: Blackwell Scientific Publications, 1983).

6   T. L. Erwin, "How Many Species Are There?: Revisited," *Conservation Biology* 5, no. 3 (1991): 330–33.

7   C. Mora, D. P. Tittensor, S. Adl, A. G. B. Simpson, and B. Worm, "How Many Species Are There on Earth and in the Ocean?," PLOS *Biology* 9, no. 8 (2011), doi:10.1371/journal.pbio.1001127.

8   R. May, "Tropical Arthropod Species, More or Less?," *Science* 329 (2010): 41–42.

9   B. Hölldobler and E. O. Wilson, *The Ants* (Cambridge, Massachusetts: Harvard University Press, 1990).

10   A. J. Hamilton, Y. Basset, K. K. Benke, P. S. Grimbacher, S. E. Miller, V. Novotny, G. A. Samuelson, N. E. Storke, G. D. Weiblen, and J. D. Yen, "Quantifying Uncertainty in Estimation of Tropical Arthropod Species Richness," *American Naturalist* 176, no. 1 (2010): 90–95.

11   H. ter Steege et al., "Hyperdominance in the Amazonian Tree Flora," *Science* 342 (2013): 1–9, doi:10.1126/science.1243092.

12   K. R. Lips, F. Brem, R. Brenes, J. D. Reeve, R. A. Alford, J. Voyles, C. Carey, L. Livo, A. P. Pessier, and J. P. Collins, "Emerging Infectious Disease and the Loss of Biodiversity in a Neotropical Amphibian Community," *Proceedings of the National Academy of Sciences* 103, no. 9 (2006): 3165–70; T. L. Chen, S. M. Rovito, D. B. Wake, and V. T. Vredenburg, "Coincident Mass Extirpation of Neotropical Amphibians with the Emergence of the Infectious Fungal Pathogen Batrachochytrium dendrobatidis," *Proceedings of the National Academy of Sciences* 108, no. 23 (2011): 9502–07.

13   Millennium Ecosystem Assessment, *Ecosystems and Human Well-Being: Synthesis* (Washington, DC: Island Press, 2005).

14   E. O. Wilson, ed., *Biodiversity,* (Washington, DC: National Academy Press, 1988).

15   N. Myers, "Tropical Forests and Their Species: Going, Going?," in *Biodiversity,* ed. E. O. Wilson (Washington, DC: National Academy Press, 1988), 28–35.

16   A. E. Lugo, "Estimating Reductions in the Diversity of Tropical Forest Species," in *Biodiversity,* ed. E. O. Wilson (Washington, DC: National Academy Press, 1988), 58–70.

17   S. Hecht and S. Saatchi, "Globalization and Forest Resurgence: Changes in Forest Cover in El Salvador," *Bioscience* 57, no. 8 (2007): 663–72.

18   O. Komar, "Avian Diversity in El Salvador," *Wilson Bulletin* 110, no. 4 (1998): 511–33.

19   B. Groombridge and M. D. Jenkins, *World Atlas of Biodiversity* (Cambridge, UK: UNEP-WCMC, 2002).

20   *The Economist,* "Special Report on Biodiversity," 12 September, 2013.

21   L. Braat and P. ten Brink, eds., "The Cost of Policy Inaction: The Case of Not Meeting the 2010 Biodiversity Target," Report for the European Commission

(Wageningen/Brussels, 2008), http://ec.europa.eu/environment/nature/biodiversity/economics/pdf/copi.zip.

22 R. McLellan, L. Iyengar, B. Jeffries, and N. Oerlemans, eds., *Living Planet Report* (Gland, Switzerland: WWF International, 2014).

23 R. McLellan, L. Iyengar, B. Jeffries, and N. Oerlemans, eds., *Living Planet Report* (Gland, Switzerland: WWF International, 2014).

24 R. H. MacArthur and E. O. Wilson, *The Theory of Island Biogeography* (Princeton, New Jersey: Princeton University Press, 1967).

25 J. B. Losos and R. E. Ricklets, eds., *The Theory of Island Biogeography Revisited* (Princeton, New Jersey: Princeton University Press, 2010).

26 R. H. MacArthur, *Geographical Ecology: Patterns in the Distribution of Species* (New York: Harper & Row, 1972).

27 W. F. Laurance, "Theory Meets Reality: How Habitat Fragmentation Research Has Transcended Island Biogeographic Theory," *Biological Conservation* 141 (2008): 1731–44.

28 W. F. Laurance, J. L. C. Camargo, R. C. C. Luizão, S. G. Laurance, S. L. Pimm, E. M. Bruna, P. C. Stouffer, G. B. Williamson, J. Benítez-Malvido, H. L. Vasconcelos, K. S. Van Houtan, C. E. Zartman, S. A. Boyle, R. K. Didham, A. Andrade, and T. E. Lovejoy, "The Fate of Amazonian Forest Fragments: A 32-Year Investigation," *Biological Conservation* 144 (2011): 56–67.

29 B. A. Wilcox and D. D. Murphy, "Conservation Strategy—Effects of Fragmentation on Extinction," *American Naturalist* 125 (1985): 879–87.

30 P. C. Stouffer, E. I. Johnson, R. O. Bierregaard Jr., and T. E. Lovejoy, "Understory Bird Communities in Amazonian Rainforest Fragments: Species Turnover through 25 Years Post-Isolation in Recovering Landscapes," PLOS ONE 6, no. 6 (2011): e20543, doi:10.1371/journal.pone.0020543.

31 P. C. Stouffer, E. I. Johnson, R. O. Bierregaard Jr., and T. E. Lovejoy, "Understory Bird Communities in Amazonian Rainforest Fragments: Species Turnover through 25 Years Post-Isolation in Recovering Landscapes," PLOS ONE 6, no. 6 (2011): e20543, doi:10.1371/journal.pone.0020543.

32 W. F. Laurance, J. L. C. Camargo, R. C. C. Luizão, S. G. Laurance, S. L. Pimm, E. M. Bruna, P. C. Stouffer, G. B. Williamson, J. Benítez-Malvido, H. L. Vasconcelos, K. S. Van Houtan, C. E. Zartman, S. A. Boyle, R. K. Didham, A. Andrade, and T. E. Lovejoy, "The Fate of Amazonian Forest Fragments: A 32-Year Investigation," *Biological Conservation* 144 (2011): 56–67.

33 A. E. Lugo, "Estimating Reductions in the Diversity of tropical Forest Species," in *Biodiversity*, ed. E. O. Wilson (Washington, DC: National Academy Press, 1988),

58–70; S. Hecht and S. Saatchi, "Globalization and Forest Resurgence: Changes in Forest Cover in El Salvador," *Bioscience* 57, no. 8 (2007): 663–72.

34   C. Martin, *The Rainforests of West Africa: Ecology—Threats—Conservation* (Basel, Boston, Berlin: Birkhäuser, 1991).

35   T. T. Struhsaker, "Conservation of Red Colobus and Their Habitats," *International Journal of Primatology* 26, no. 3 (2005): 525–38.

36   E. D. Wiafe, "Status of the Critically Endangered Roloway Monkey (*Cercopithecus diana roloway*) in Dadieso Forest Reserve, Ghana," *African Primates* 8 (2013): 9–16.

37   E. O. Effiom, G. Nuñez-Iturri, H. G. Smith, U. Ottosson, and O. Olsson, "Bushmeat Hunting Changes Regeneration of African Rainforests," *Proceedings of the Royal Society B* 280, no. 1759 (2013): 20130246, doi:10.1098/rspb.2013.0246.

38   K. H. Redford, "The Empty Forest," *BioScience* 42, no. 6 (1992): 412–22.

39   D. J. McCauley, "Selling Out on Nature," *Nature* 443 (2006): 27–28.

40   N. Stern, *The Economics of Climate Change: The Stern Review* (Cambridge, UK: Cambridge University Press, 2007).

41   P. Sukhdev, "Will 2013 Bring a New, Sustainable World?," 5 January, 2013, http://pavansukhdev.com/?m=20130105.

42   L. Braat and P. ten Brink, eds., "The Cost of Policy Inaction: The Case of Not Meeting the 2010 Biodiversity Target," Report for the European Commission (Wageningen/Brussels, 2008), http://ec.europa.eu/environment/nature/biodiversity/economics/pdf/copi.zip.

43   TEEB for Business Coalition/Trucost, "Natural Capital at Risk: The Top 100 Externalities of Business," 2013, http://www.trucost.com/published-research/99/natural-capital-at-risk-the-top-100-externalities-of-business.

44   C. L. Spash, "Terrible Economics, Ecosystems and Banking," *Environmental Values* 20 (2011): 141–45.

45   *The Economist*, "Special Report on Biodiversity," 12 September, 2013.

46   C. Zhu, R. Taylor, and G. Feng, *China's Wood Market, Trade and the Environment* (Monmouth Junction, New Jersey: Science Press USA, 2004).

47   C. L. Spash, "Terrible Economics, Ecosystems and Banking," *Environmental Values* 20 (2011): 141–45.

SPECIALIST'S VIEW: KAIMOWITZ

1   Rights and Resources Initiative, *What Future for Reform? Progress and Slowdown in Forest Tenure Reform Since 2002* (Washington D.C: Rights and Resources Initiative, 2014).

2   C. Van Dan, "Indigenous Territories and REDD in Latin America: Opportunity or Threat?," *Forests* 2 (2011): 394–414.

3   Rights and Resources Initiative, *What Future for Reform? Progress and Slowdown in Forest Tenure Reform Since 2002* (Washington D.C: Rights and Resources Initiative, 2014).

4   G. Vergara Asenjo and C. Potvin, "Forest Protection and Tenure Status: The Key Role of Indigenous Peoples and Protected Areas in Panama," *Global Environmental Change* 28 (2014): 205–15, doi:10.1016/j.gloenvcha.2014.07.002.

5   Rights and Resources Initiative, *What Future for Reform? Progress and Slowdown in Forest Tenure Reform Since 2002* (Washington D.C: Rights and Resources Initiative, 2014).

6   D. Armenteras, N. Rodriguez, and J. Retana, "Are Conservation Strategies Effective in Avoiding the Deforestation of the Colombian Guyana Shield?," *Biological Conservation* 142 (2009): 1411–1419; M. Bonilla-Moheno, D. Redo, T. M. Aide, M. Clark, and H. R. Grau, "Vegetation Change and Land Tenure in Mexico: A Country-Wide Analysis," *Land Use Policy* 30 (2013): 355–64; T. Hayes, "Controlling Agricultural Expansion in the Mosquitia: Does Tenure Matter?," *Human Ecology* 35, no. 6 (2007): 733–47; T. J. Killeen, A. Guerra, M. Calzada, L. Correa, V. Calderon, L. Soria, B. Quezada, and M. K. Steininger, "Total Historical Land-Use Change in Eastern Bolivia: Who, Where, When, and How Much?," *Ecology and Society* 13, no. 1 (2008): 36, http://www.ecologyandsociety.org/vol13/iss1/art36/; D. Nepstad, S. Schwartz-man, B. Bamberger, M. Santilli, D. Ray, P. Schlesinger, P. Lefebvre, A. Alencar, E. Prinz, G. Fiske, and A. Rolla, "Inhibition of Amazon Deforestation and Fire by Parks and Indigenous Lands," *Conservation Biology* 20 (2006): 65–73; P. Oliveira, G. Asner, D. Knap, A. Almeyda, R. Galván-Gildemeister, S. Keene, R. Raybin, and R. Smith, "Land-Use Allocation Protects the Peruvian Amazon," *Science* 317, no. 5842 (2007): 1233–36; T. H. Ricketts, B. Soares Filho, G. A. B. da Fonseca, D. Nepstad, A. Pfaff, A. Petsonk, A. Anderson, D. Boucher, A. Cattaneo, M. Conte, K. Creighton, L. Linden, C. Maretti, P. Moutinho, R. Ullman, and R. Victurine, "Indigenous Lands, Protected Areas, and Slowing Climate Change," PLOS *Biology* 8, no. 3 (2010), doi:10.1371/journal.pbio.1000331; A. Stocks, B. McMahan, and P. Taber, "Indigenous, Colonist, and Government Impacts on Nicaragua's Bosawas Biosphere Reserve," *Conservation Biology* 21, no. 6 (2007): 1495–1505; G. Vergara Asenjo and C. Potvin, "Forest Protection and Tenure Status: The Key Role of Indigenous Peoples and Protected Areas in Panama," *Global Environmental Change* 28 (2014): 205–15, doi:10.1016/j.gloenvcha.2014.07.002.

7   M. B. Holland, F. de Koning, M. Morales, L. Naughton Treves, B. E. Robinson, and L. Suárez, "Complex Tenure and Deforestation: Implications for Conservation Incentives in the Ecuadorian Amazon," *World Development* 55 (2014): 21–36.

8   A. Nelson and K. Chomitz, *Protected Area Effectiveness in Reducing Tropical Defor-estation: A Global Analysis of the Impact of Protection Status*, Evaluation Brief 7 (Washington DC: Independent Evaluation Group World Bank, 2009).

9   D. Armenteras, N. Rodriguez, and J. Retana, "Are Conservation Strategies Effec-tive in Avoiding the Deforestation of the Colombian Guyana Shield?," *Biological Conservation* 142 (2009): 1411–1419; P. Oliveira, G. Asner, D. Knap, A. Almeyda, R. Galván-Gildemeister, S. Keene, R. Raybin, and R. Smith, "Land-Use Allocation Protects the Peruvian Amazon," *Science* 317, no. 5842 (2007): 1233–36.

10   C. Nolte, A. Agrawal, K. M. Silvius, and B. S. Soares Filho, "Governance Regime and Location Influence: Avoided Deforestation Success in Protected Areas in the Brazilian Amazon," *Proceedings of the National Academy of Sciences* 110, no. 13 (2013): 4956–61.

11   M. B. Holland, F. de Koning, M. Morales, L. Naughton Treves, B. E. Robinson, and L. Suárez, "Complex Tenure and Deforestation: Implications for Conserva-tion Incentives in the Ecuadorian Amazon," *World Development* 55 (2014): 21–36.

12   T. K. Rudel, *Tropical Forests: Regional Paths of Destruction and Regeneration in the Late Twentieth Century* (New York: Columbia University Press, 2013).

13   A. Angelsen and D. Kaimowitz, *Agricultural Technologies and Tropical Deforestation* (Wallingford, UK: CABI Publishing, 2011).

14   A. Nelson and K. Chomitz, *Protected Area Effectiveness in Reducing Tropical Defor-estation: A Global Analysis of the Impact of Protection Status*, Evaluation Brief 7 (Washington DC: Independent Evaluation Group World Bank, 2009); C. Nolte, A. Agrawal, K. M. Silvius, and B. S. Soares Filho, "Governance Regime and Location Influence: Avoided Deforestation Success in Protected Areas in the Brazilian Amazon," *Proceedings of the National Academy of Sciences* 110, no. 13 (2013): 4956–61.

15   J. Alcorn and V. M. Toledo, "Resilient Resource Management in Mexico's Forest Ecosystems: The Contribution of Property Rights," in F. Berkes, C. Folke, and J. Colding, *Linking Social and Ecological Systems: Management Practices and Social Mecha-nisms for Building Resilience* (New York: Columbia University Press, 2000), 216–30; C. L. Gray, R. E. Bilsborrow, J. L. Bremmer, and F. Lu, "Indigenous Land Use in the Ecuadorian Amazon: A Cross-Cultural and Multilevel Analysis," *Human Ecology* 36, no. 1 (2008): 97–109; T. J. Killeen, A. Guerra, M. Calzada, L. Correa, V. Calderon, L. Soria, B. Quezada, and M. K. Steininger, "Total Historical Land-Use Change in Eastern Bolivia: Who, Where, When, and How Much?," *Ecology and Society* 13, no. 1 (2008): 36, http://www.ecologyandsociety.org/vol13/iss1/art36/; T. K. Rudel, D. Bates, and R. Machinguiashi, "Ecologically Noble Amerindians,"

Latin American Research Review 37, no. 1 (2002): 144–59; A. Stocks, B. McMahan, and P. Taber, "Indigenous, Colonist, and Government Impacts on Nicaragua's Bosawas Biosphere Reserve," *Conservation Biology* 21, no. 6 (2007): 1495–1505.

16   Rights and Resources Initiative, *The End of the Hinterland: Forests, Conflict, and Climate Change* (Washington DC: Rights and Resources Initiative, 2010).

CHAPTER 7

1   C. Martin, *The Rainforests of West Africa: Ecology—Threats—Conservation* (Basel, Boston, Berlin: Birkhäuser, 1991).

2   M. Williams, *Deforesting the Earth: From Prehistory to Global Crisis* (Chicago and London: University of Chicago Press, 2003).

3   P. W. Richards, *The Tropical Rain Forest* (London: Cambridge University Press, 1952).

4   A. Sommer, "Attempt at an Assessment of the World's Tropical Moist Forests," *Unasylva* 28, no. 112–13 (1976): 5–25.

5   N. Myers, *The Sinking Ark: A New Look at the Problem of Disappearing Species* (Oxford: Pergamon Press, 1979).

6   N. Myers, *Conversion of Tropical Moist Forests: A Report Prepared for the Committee on Research Priorities in Tropical Biology of the National Research Council* (Washington, DC: National Academy of Sciences, 1980); N. Myers, *A Wealth of Wild Species* (Boulder, Colorado: Westview Press, 1983).

7   E. O. Wilson, ed., *Biodiversity*, (Washington, DC: National Academy Press, 1988).

8   J. Sayer, "Conservation and Protection of Tropical Rain Forests: The Perspective of the World Conservation Union–IUCN," *Unasylva* 166 (1991): 40–45.

9   N. Dudley and A. Phillips, *Forest and Protected Areas: Guidance on the Use of the IUCN Protected Area Management Categories*, World Commission on Protected Areas (WCPA) Best Practice Protected Area Guidelines Series No. 12 (Gland, Switzerland: IUCN-The World Conservation Union, 2006).

10   S. Chape, M. Spalding, and M. D. Jenkins, *The World's Protected Areas*, World Conservation Monitoring Centre (WCMC) (Berkeley, California: University of California Press, 2008).

11   C. B. Schmitt, A. Belokurov, C. Besanc̦on, L. Boisrobert, N. D. Burgess, A. Campbell, L. Coad, L. Fish, D. Gliddon, K. Humphries, V. Kapos, C. Loucks, I. Lysenko, L. Miles, C. Mills, S. Minnemeyer, T. Pistorius, C. Ravilious, M. Steininger, and G. Winkel, *Global Ecological Forest Classification and Forest Protected Area Gap Analysis: Analyses and Recommendations in View of the 10% Target for Forest Protection under the*

Convention on Biological Diversity (CBD) (Freiburg, Germany: Freiburg University Press, 2008).

12   S. Chape, S. Blythe, L. Fish, P. Fox, and M. Spalding, compilers, 2003 United Nations List of Protected Areas (Gland, Switzerland: IUCN and Cambridge, UK: UNEP World Conservation Monitoring Centre, 2003).

13   C. B. Schmitt, A. Belokurov, C. Besançon, L. Boisrobert, N. D. Burgess, A. Campbell, L. Coad, L. Fish, D. Gliddon, K. Humphries, V. Kapos, C. Loucks, I. Lysenko, L. Miles, C. Mills, S. Minnemeyer, T. Pistorius, C. Ravilious, M. Steininger, and G. Winkel, Global Ecological Forest Classification and Forest Protected Area Gap Analysis: Analyses and Recommendations in View of the 10 Percent Target for Forest Protection under the Convention on Biological Diversity (CBD) (Freiburg, Germany: Freiburg University Press, 2008).

14   IUCN, UNEP, and WWF, Caring for the Earth–A Strategy for Sustainable Living (Gland, Switzerland: IUCN, 1991).

15   M. E. Soulé and M. A. Sanjayan, "Conservation Targets: Do They Help?," Science 279 (1998): 2060–61.

16   E. Nicholson, "Testing the Waters–From Global CBD Targets to Indicators to Achievements," Decision Point 64 (2012): 10–11.

17   S. Chape, M. Spalding, and M. D. Jenkins, The World's Protected Areas, World Conservation Monitoring Centre (WCMC) (Berkeley, California: University of California Press, 2008).

18   Congo Basin Forest Partnership, http://pfbc-cbfp.org/home.html.

19   WWF, "The Heart of Borneo Declaration," http://wwf.panda.org/what_we_do/where_we_work/borneo_forests/about_borneo_forests/declaration.cfm.

20   WWF, Borneo: Treasure Island at Risk (Frankfurt: WWF Germany, 2005), http://d2ouvy59p0dg6k.cloudfront.net/downloads/treasureislandatrisk.pdf.

21   S. Wulffraat, 2012. The Environmental Status of the Heart of Borneo (WWF, 2012), http://d2ouvy59p0dg6k.cloudfront.net/downloads/wwf___hob_measures_report___2012___final_for_web.pdf.

22   A. Langner, J. Miettinen, and F. Siegert, "Land Cover Change 2002–2005 in Borneo and the Role of Fire Derived from MODIS Imagery," Global Change Biology 13 (2007): 2329–40.

23   S. Wulffraat, 2012. The Environmental Status of the Heart of Borneo (WWF, 2012), http://d2ouvy59p0dg6k.cloudfront.net/downloads/wwf___hob_measures_report___2012___final_for_web.pdf.

24   S. Wulffraat, 2012. The Environmental Status of the Heart of Borneo (WWF, 2012), http://d2ouvy59p0dg6k.cloudfront.net/downloads/wwf___hob_measures_report___2012___final_for_web.pdf.

25   Food and Agriculture Organization, *Global Forest Resources Assessment* 2010: *Main Report*, FAO Forestry Paper 163 (Rome: FAO, 2010).

26   K. J. Willis and G. M. MacDonald, "Long-Term Ecological Records and Their Relevance to Climate Change Predictions for a Warmer World," *Annual Review of Ecology, Evolution, and Systematics* 42 (2011): 267–87.

27   L. Gibson, T. M. Lee, L. P. Koh, B. W. Brook, T. A. Gardner, J. Barlow, C. A. Peres, C. J. A. Bradshaw, W. L. Laurance, T. E. Lovejoy, and N. S. Sodhi, "Primary Forests Are Irreplaceable for Sustaining Tropical Biodiversity," *Nature* 478 (2011): 378–81, doi:10.1038/nature10425; C. J. Clark, J. R. Poulsen, R. Malonga, and P. W. Elkan, "Logging Concessions Can Extend the Conservation Estate for Central African Tropical Forests," *Conservation Biology* 23 (2009): 1281–93, doi:10.1111/j.1523-1739.2009.01243.x.

28   P. Potapov, A. Yaroshenko, S. Turubanova, M. Dubinin, L. Laestadius, C. Thies, D. Aksenov, A. Egorov, Y. Yesipova, I. Glushkov, M. Karpachevskiy, A. Kostikova, A. Manisha, E. Tsybikova, and I. Zhuravleva, "Mapping the World's Intact Forest Landscapes by Remote Sensing," Ecology and Society 13, no. 2 (2008): 51, http://www.ecologyandsociety.org/vol13/iss2/art51.

29   P. Potapov, A. Yaroshenko, S. Turubanova, M. Dubinin, L. Laestadius, C. Thies, D. Aksenov, A. Egorov, Y. Yesipova, I. Glushkov, M. Karpachevskiy, A. Kostikova, A. Manisha, E. Tsybikova, and I. Zhuravleva, "Mapping the World's Intact Forest Landscapes by Remote Sensing," *Ecology and Society* 13, no. 2 (2008): 51, http://www.ecologyandsociety.org/vol13/iss2/art51.

30   C. Nolte, A. Agrawal, K. M. Silvius, and B. S. Soares Filho, "Governance Regime and Location Influence Avoided Deforestation Success of Protected Areas in the Brazilian Amazon," *Proceedings of the National Academy of Sciences* 110 (2013): 4956–61, doi:10.1073/pnas.1214786110.

SPECIALIST'S VIEW: MARETTI

1   I am grateful for the collaboration, revision, and comments by Claude Martin, Alejandro Coca-Castro, Louis Reymondin, Denise Oliveira, André S. Dias, Juan Carlos Riveros S., Robert Hofstede, and others. This contribution is partly based on internal WWF reports, particularly Riveros et al. 2014 and my own assessments.

2   WWF, "Amazon strategy planning internal work," unpublished, 2007–08; M. Flores, U. Lopes da Silva Jr., H. Malone, M. Panuncio, J. C. Riveros, S. Rodrigues, R. Silva, S. Valenzuela, D. Arancibia, P. Bara-Neto, and M. Symington, WWF's *Living Amazon Initiative: A Comprehensive Approach to Conserving the Largest Rainforest and River System on Earth*, WWF Strategy Summary (WWF, 2010); H. D. Eva and

O. Huber, eds., *A Proposal for Defining the Geographical Boundaries of Amazonia* (European Commission, Directorate General, Joint Research Center and ACTO, 2005); RAISG, *Amazonia under Pressure*, 2012, www.raisg.socioambiental.org.

3   *SavingSpecies*, "Stunning New Biodiversity Maps Show Where to Prioritize Conservation," blog entry by C. N. Jenkins, 5 September, 2012, http://savingspecies.org/2012/stunning-new-biodiversity-maps-show-where-to-prioritize-conservation; C. N. Jenkins, S. L. Pimm, and L. N. Joppa, "Global Patterns of Terrestrial Vertebrate Diversity and Conservation," *Proceedings of the National Academy of Sciences* 110, no. 28 (2013): E2602–E2610, doi:10.1073/pnas.1302251110; RedParques, *Progress in the Development of the Program of Work on Protected Areas: Region: Amazon Biome*, Report and 10-year action plan, October 2010, https://portals.iucn.org/2012forum/sites/2012forum/files/cop-10-eng-baja.pdf; Millennium Ecosystem Assessment, *Ecosystems and Human Well-Being: Synthesis* (Washington, DC: Island Press, 2005); A. Ruesch and H. K. Gibbs, *New IPCC Tier-1 Global Biomass Carbon Map For the Year 2000* (Oak Ridge, Tennessee: Oak Ridge National Laboratory, 2008), http://cdiac.ornl.gov/epubs/ndp/global_carbon/carbon_documentation.html; J. P. W. Scharlemann, R. Hiederer, V. Kapos, and C. Ravilious, UNEP WCMC *Updated Global Carbon Map* (German Federal Agency for Nature Conservation, 2011) http://eusoils.jrc.ec.europa.eu/esdb_archive/octop/Resources/Global_OC_Poster.pdf.

4   P. H. May, B. Millikan, and M. F. Gebara, "The Context of REDD+ in Brazil: Drivers, Agents and Institutions," CIFOR Occasional Paper 55, 2nd ed. (Bogor, Indonesia: CIFOR, 2011); C. C. Maretti, L. H. O. Wadt, D. A. P. Gomes-Silva, W. T. P. Maldonado de V., R. A. Sanches, F. Coutinho, and S. da S. Brito, "From Pre-assumptions to a 'Just World Conserving Nature': The Role of Category VI in Protecting Landscapes," in *The Protected Landscape Approach: Linking Nature, Culture and Community*, eds. J. Brown, N. Mitchell, and M. Beresford (Gland, Switzerland, and Cambridge, UK: IUCN, 2005), 47–64; RAISG, *Amazonia under Pressure*, 2012, www.raisg.socioambiental.org; RedParques, *Progress in the Development of the Program of Work on Protected Areas: Region: Amazon Biome*, Report and 10-year action plan, October 2010, https://portals.iucn.org/2012forum/sites/2012forum/files/cop-10-eng-baja.pdf.

5   J. C. Riveros S., R. Hofstede, T. Granizo, C. C. Maretti, and D. Oliveira, *Protected Areas and Indigenous Territories of the Amazon—Five Decades of Change (1960–2012)*—a WWF Living Amazon Initiative report (internal WWF draft), 2014; RedParques, *Progress in the Development of the Program of Work on Protected Areas: Region: Amazon Biome*, Report and 10-year action plan, October 2010, https://portals.iucn.org/2012forum/sites/2012forum/files/cop-10-eng-baja.pdf; Infoamazonia, "World Database on Protected Areas, Amzon Countries,"

http://infoamazonia.org/datasets/sources/protected-planet-the-world-data-base-on-protected-areas-wdpa; J. Beltrán and A. Phillips, eds., *Indigenous and Traditional Peoples and Protected Areas: Principles, Guidelines and Case Studies* (Gland, Switzerland, and Cambridge, UK: IUCN, 2000); G. Borrini-F., "Indigenous and Local Communities and Protected Areas: Rethinking the Relationship," in *Parks, Local Communities and Protected Areas* 12, no. 2 (2002): 5–15; C. C. Maretti, "Conservação e valores; relações entre áreas protegidas e indígenas: possíveis conflitos e soluções," in *Terras indígenas & unidades de conservação da natureza: o desafio das sobreposições* ed. F. Ricardo (São Paulo, Brazil: Instituto Socioambiental, 2005), 85–101; C. C. Maretti, et al. "Brazil: Lessons Learned in the Establishment and Management of Protected Areas by Indigenous and Local Communities," paper presented at the World Parks Congress, Durban, 2003; G. Oviedo, "Lessons Learned in the Establishment and Management of Protected Areas by Indigenous and Local Communities," paper presented at the World Parks Congress, Durban 2003.

6   Among the large national parks created in the 1970s were Manu, Yasuní, Jaú, and Pico da Neblina. By 2010, there were 102 protected areas covering more than half a million hectares.

7   WWF in 1998, through its then director general Claude Martin, challenged the Brazilian president Fernando Henrique Cardoso to increase the protected areas coverage of the Amazon to a minimum of 10 percent. This target was in line with the World Bank–WWF Alliance for Forest Conservation and Sustainable Use. After the agreement was signed in 2002, the program evolved and the target was increased to fifty million hectares (12 percent of the Brazilian Amazon). As an important part of the target was achieved in the course of its first phase (mostly under President "Lula" da Silva), the ARPA target was further increased to sixty million hectares in its second phase. In the year 2002 alone, fourteen protected areas were created in Brazil, including the iconic "Montanhas do Tumucumaque" National Park, an area as large as the Netherlands.

8   Significant parts of these mosaics were linked to the ARPA program; others were supported by nongovernmental organizations such as Conservation International, Greenpeace, and the Brazilian Instituto Socioambiental ISA and Instituto Centro da Vida ICV.

9   C. N. Jenkins and L. Joppa, "Expansion of the Global Terrestrial Protected Area System," *Biological Conservation* 142, no. 10 (2009): 2166–74, doi:10.1016/ j.biocon.2009.04.016; The most impressive increase happened in French Guiana with the creation in 2007 of the Parc amazonien de Guyane. In absolute numbers Brazil has created most new protected areas in the last decade. Jenkins and

Joppa (previous citation in this note) mention that Brazil is responsible for 86% of the global increase in protected areas (with IUCN categories) between 2003 and 2009, obviously related to its protect areas in the Amazon.

10   J. C. Riveros S., R. Hofstede, T. Granizo, C. C. Maretti, and D. Oliveira, Protected Areas and Indigenous Territories of the Amazon—Five Decades of Change (1960–2012)—a WWF Living Amazon Initiative report (internal WWF draft), 2014.

11   C. C. Maretti, M. I. S. Catapan, M. J. P. de Abreu, J. E. D. de Oliveira, "Áreas protegidas: definições, tipos e conjuntos. Reflexões conceituais e diretrizes para gestão," in *Gestão de Unidades de Conservação: compartilhando uma experiência de capacitação*, ed. M. O. Cases (Brasília, Brazil: WWF-Brazil and IPÊ Instituto de Pesquisas Ecológicas, 2012), 331–67; J. C. Riveros S., R. Hofstede, T. Granizo, C. C. Maretti, and D. Oliveira, Protected Areas and Indigenous Territories of the Amazon—Five Decades of Change (1960–2012)—a WWF Living Amazon Initiative report (internal WWF draft), 2014.

12   The two groups of Brazilian category six protected areas are: 1) extractive reserves and sustainable development reserves, with some kind of comanagement with local communities (about eighty in number covering twenty-four million hectares); 2) the national, state, and municipal forests, with objectives linked to forestry (fifty-eight in number covering thirty-three million hectares). Areas in the latter category can only be considered protected areas if they are designed and managed according to the objectives of the National System of Protected Areas (Sistema Nacional de Unidades de Conservação—SNUC) legislation.

13   G. Borrini-F., "Indigenous and Local Communities and Protected Areas: Rethinking the Relationship," in *Parks, Local Communities and Protected Areas* 12, no. 2 (2002): 5–15; A. Kothari, C. Corrigan, H. Jonas, A. Neumann, and H. Shrumm, eds., *Recognising and Supporting Territories and Areas Conserved By Indigenous Peoples and Local Communities: Global Overview and National Case Studies*, Technical Series 64 (Secretariat of the Convention of Biological Diversity, 2012); C. C. Maretti et al. "Brazil: Lessons Learned in the Establishment and Management of Protected Areas by Indigenous and Local Communities," paper presented at the World Parks Congress, Durban, 2003; C. C. Maretti, M. I. S. Catapan, M. J. P. de Abreu, J. E. D. de Oliveira, "Áreas protegidas: definições, tipos e conjuntos. Reflexões conceituais e diretrizes para gestão," in *Gestão de Unidades de Conservação: compartilhando uma experiência de capacitação*, ed. M. O. Cases (Brasília, Brazil: WWF-Brazil and IPÊ Instituto de Pesquisas Ecológicas, 2012), 331–67; M. R. Pinheiro, C. Delelis, C. Costa, C. F. Lino, H. Dias, I. V. Ferreira, I. R. Lamas, M. R. Lederman, R. V. Fernandes, and T. M. Cardoso, *Recomendações para reconhecimento e implementação de mosaicos de áreas protegidas* (Brasília, Brazil: GTZ, 2010).

14   Most information about indigenous territories is based on the following two
RAISG papers and the interpretation by Riveros et al. cited here: RAISG, Raw
data on Amazon indigenous territories collected from several sources (unpublished spreadsheet, 2010); RAISG, *Amazonia under Pressure*, 2012, www.raisg.
socioambiental.org.; J. C. Riveros S., R. Hofstede, T. Granizo, C. C. Maretti, and
D. Oliveira, Protected Areas and Indigenous Territories of the Amazon—Five
Decades of Change (1960–2012)—a WWF Living Amazon Initiative report
(internal WWF draft), 2014. *Amazonia under Pressure* gives the total area as 214
million hectares, of which some 77 percent were officially recognized by the
various Amazon governments. A further 20 percent of the Amazon indigenous
territories were not yet officially recognized or were without information, and
3 percent was classified under other definitions. Also, more than 33.6 million
hectares of indigenous territories overlapped protected areas designated for
nature conservation.

15   RAISG, *Amazonia under Pressure*, 2012, www.raisg.socioambiental.org.

16   Most of the discussion of ecological representation in the Amazon and the
information about the current status is based on Riveros et al. 2014. These target
values are assumed to vary in response to various factors in a region or habitat
type, including connectivity, natural disturbances, and human resource uses.

17   N. Dudley, ed., *Guidelines for Applying Protected Area Management Categories,* (Gland,
Switzerland: IUCN, 2008); M. B. Mascia and S. Pailler, "Protected Area Down-
grading, Downsizing, and Degazettement (PADDD) and Its Conservation
Implications," *Conservation Letters* 4 (2011): 9–20; M. B. Mascia, S. Pailler, R.
Krithivasan, V. Roshchanka, D. Burns, M. J. Mlotha, D. R. Murray, and N. Peng,
"Protected Area Downgrading, Downsizing, and Degazettement (PADDD) in
Africa, Asia, and Latin America and the Caribbean, 1900–2010," *Biological Conser-
vation* 169 (2014): 355–61; H. Martins, M. Vedoveto, E. Araújo, P. Barreto, S. Baima,
C. Souza Jr., and A. Veríssimo, *Áreas protegidas críticas na Amazônia legal* (Belém, Bra-
zil: Imazon, 2012).

18   D. Nepstad, S. Schwartzman, B. Bamberger, M. Santilli, D. Ray, P. Schlesinger,
P. Lefebvre, A. Alencar, E. Prinz, G. Fiske, and A. Rolla, "Inhibition of Amazon
Deforestation and Fire by Parks and Indigenous Lands," *Conservation Biology* 20
(2006): 65–73; B. Soares Filho, L. Dietzsch, P. Moutinho, A. Falieri, H. Rodrigues,
E. Pinto, C. C. Maretti, C. A. Scaramuzza de M., A. Anderson, K. Suassuna, M.
Lanna, and F. Vasconcelos de Araújo, *Reducing Carbon Emissions from Defor-
estation: The Role of ARPA's Protected Areas in the Brazilian Amazon* (Brasília, Brazil:
WWF Brazil, 2009); B. Soares Filho, P. Moutinho, D. Nepstad, M. Bowman, H.
Rodrigues, A. Anderson, R. Garcia, L. Dietzsch, F. Merry, L. Hissa, R. Silves-
trini, and C. C. Maretti, "Role of Brazilian Amazon Protected Areas in Climate

Change Mitigation," *Proceedings of the National Academy of Sciences* 107, no. 24 (2010): 10821–26; T. H. Ricketts, B. Soares Filho, G. A. B. da Fonseca, D. Nepstad, A. Pfaff, A. Petsonk, A. Anderson, D. Boucher, A. Cattaneo, M. Conte, K. Creighton, L. Linden, C. C. Maretti, P. Moutinho, R. Ullman, R. Victurine, "Indigenous Lands, Protected Areas, and Slowing Climate Change," PLOS *Biology* 8, no. 3 (2010): e1000331, doi:10.1371/journal.pbio.1000331; C. Nolte, A. Agrawala, K. M. Silvius, and B. Soares Filho, "Governance Regime and Location Influence Avoided Deforestation Success of Protected Areas in the Brazilian Amazon," *Proceedings of the National Academy of Sciences* 110, no. 13 (2013): 4956–61, doi:10.1073/pnas.1214786110.

19   Deforestation pressures may in the future become less related to frontier activities such as land speculation and grabbing, low-productivity cattle ranching, illegal logging, etc., and shift to capital-intensive high-technology agriculture, hydropower, oil and gas exploitation, mining, and so on. Protected areas will be less likely to keep their important role under such circumstances. Negotiated approaches that include protected areas as mitigation and compensation or off-setting tools may be necessary. Protected areas may even be partially funded by some of the businesses interests involved.

CHAPTER 8

1   O. Edenhofer, R. Pichs-Madruga, Y. Sokona, E. Farahani, S. Kadner, K. Seyboth, A. Adler, I. Baum, S. Brunner, P. Eickemeier, B. Kriemann, J. Savolainen, S. Schlömer, C. von Stechow, T. Zwickel, and J. C. Minx, eds., *Climate Change 2014: Mitigation of Climate Change. Contribution of Working Group III to the Fifth Assessment Report of the Intergovernmental Panel on Climate Change* (Cambridge, UK, and New York: Cambridge University Press, 2014); Union of Concerned Scientists blog, *The Equation,* "10% of Greenhouse Gas Emissions Come from Deforestation," blog entry by D. Boucher, 12 December, 2013, http://blog.ucsusa.org/ten-percent-of-greenhouse-gas-emissions-come-from-deforestation-342.

2   R. Bonnefille, "Rainforest Responses to Past Climatic Changes in Tropical Africa," in *Tropical Rainforest Responses to Climatic Change,* 2nd ed., M. B. Bush, J. R. Flenley, and W. D. Gosling, eds. (Berlin: Springer, 2011), 125–84; J. Maley, "The African Rain Forest: Main Characteristics of Changes in Vegetation and Climate from the Upper Cretaceous to the Quaternary," *Proceedings of the Royal Society of Edinburgh, Section B* 104 (1996): 31–73.

3   K. J. Willis and G. M. MacDonald, "Long-Term Ecological Records and Their Relevance to Climate Change Predictions for a Warmer World," *Annual Review of Ecology, Evolution, and Systematics* 42 (2011): 267–87; K. J. Willis, K. D. Bennett, S. L.

Burrough, M. Macias-Fauria, and C. Tovar, "Determining the Response of African Biota to Climate Change: Using the Past to Model the Future," *Philosophical Transactions of the Royal Society B* 368 (2013): 20120491; Y. Malhi, S. Adu-Bredu, R. A. Asare, S. L. Lewis, and P. Mayaux, "African Rainforests: Past, Present and Future," *Philosophical Transactions of the Royal Society B* 368 (2013): 20120312.

4   J. E. Nichol, "Geomorphological Evidence and Pleistocene Refugia in Africa," *Geographical Journal* 165 (1999): 79–89.

5   D. A. Livingstone, "Quaternary Geography of Africa and the Refuge Theory," in *Biological Diversification in the Tropics*, G. T. Prance, ed. (New York: Columbia University Press, 1982), 523–36.

6   A. C. Hamilton, "The Significance of Patterns of Distribution Shown by Forest Plants and Animals in Tropical Africa for the Reconstruction of Upper Pleistocene Palaeoenvironments: A Review," in *Palaeoecology of Africa & of the Surrounding Islands and Antarctica*, E. M. van Zinderen-Bakker, vol. 9 (Rotterdam, Netherlands: Balkema, 1976), 63–97.

7   R. J. Morley, "Cretaceous and Tertiary Climate Change and the Past Distribution of Megathermal Rainforests," in *Tropical Rainforest Responses to Climatic Change*, 2nd ed., M. B. Bush, J. R. Flenley, and W. D. Gosling, eds. (Berlin: Springer, 2011), 1–34.

8   P. Grubb, "Refuges and Dispersal in the Speciation of African Forest Mammals," in *Biological Diversification in the Tropics*, G. T. Prance, ed. (New York: Columbia University Press, 1982), 537–53.

9   C. Martin, *The Rainforests of West Africa: Ecology—Threats—Conservation* (Basel, Boston, Berlin: Birkhäuser, 1991).

10   N. A. Drake, R. M. Blench, S. J. Armitage, C. S. Bristow, and K. H. White, "Ancient Watercourses and Biogeography of the Sahara Explain the Peopling of the Desert," *Proceedings of the National Academy of Sciences* 108, no. 2 (2011): 458–62, doi:10.1073/pnas.1012231108.

11   K. J. Willis, K. D. Bennett, S. L. Burrough, M. Macias-Fauria, and C. Tovar, "Determining the Response of African Biota to Climate Change: Using the Past to Model the Future," *Philosophical Transactions of the Royal Society B* 368 (2013): 20120491.

12   J. P. Steffensen, et al., "High-Resolution Greenland Ice Core Data Show Abrupt Climate Change Happens in a Few Years," *Science* 321 (2008): 680–84, doi:10.1126/science.1157707.

13   Y. Malhi, S. Adu-Bredu, R. A. Asare, S. L. Lewis, and P. Mayaux, "African Rainforests: Past, Present and Future," *Philosophical Transactions of the Royal Society B* 368 (2013): 20120312.

14 P. Zelazowsky, Y. Malhi, C. Huntingford, S. Sitch, and J. B. Fisher, "Changes in the Potential Distribution of Humid Tropical Forests on a Warmer Planet," *Philosophical Transactions of the Royal Society A*: 369, no. 1934 (2011): 137–60.

15 M. I. Bird, D. Taylor, and C. Hunt, "Palaeoenvironments of Insular Southeast Asia During the Last Glacial Period: A Savanna Corridor in Sundaland?," *Quaternary Science Reviews* 24 (2005): 2228–42.

16 R. Dennell and M. D. Petraglia, "The Dispersal of *Homo sapiens* across Southern Asia: How Early, How Often, How Complex," *Quaternary Science Reviews* 47 (2012): 15–22.

17 A. P. Kershaw, S. van der Kaars, and J. R. Flenley, "The Quaternary History of Far Eastern Rainforests," in *Tropical Rainforest Responses to Climatic Change*, 2nd ed., M. B. Bush, J. R. Flenley, and W. D. Gosling, eds. (Berlin: Springer, 2011), 86–123.

18 F. E. Mayle and M. J. Power, "Impact of Drier Early-Mid Holocene Climate upon Amazonian Forests," *Philosophical Transactions of the Royal Society B*: 363, no. 1498 (2008): 1829–38.

19 Y. Malhi, S. Adu-Bredu, R. A. Asare, S. L. Lewis, and P. Mayaux, "African Rainforests: Past, Present and Future," *Philosophical Transactions of the Royal Society B* 368 (2013): 20120312.

20 National Research Council, *Abrupt Impacts of Climate Change: Anticipating Surprises*, Pre-publication version (Washington, DC: National Academies Press, 2013).

21 C. D. Allen, "Climate-Induced Forest Dieback: An Escalating Global Phenomenon?," *Unasylva* 231–232, no. 60 (2009): 43–49.

22 P. M. Cox, R. A. Betts, C. D. Jones, S. A. Spall, and I. J. Todderdell, "Acceleration of Global Warming Due to Carbon-Cycle Feedbacks in a Coupled Climate Model," *Nature* 408 (2000): 184–87.

23 S. L. Lewis, "Tropical Forests and the Changing Earth System," *Philosophical Transactions of the Royal Society B* 361 (2006): 195–210, doi:10.1098/rstb.2005.1711.

24 P. Zelazowsky, Y. Malhi, C. Huntingford, S. Sitch, and J. B. Fisher, "Changes in the Potential Distribution of Humid Tropical Forests on a Warmer Planet," *Philosophical Transactions of the Royal Society A*: 369, no. 1934 (2011): 137–60.

25 D. Nepstad, P. Lefebvre, U. L. da Silva, J. Tomasella, P. Schlesinger, L. Solórzano, P. Moutinho, D. Ray, J. G. Benito, "Amazon Drought and Its Implications for Forest Flammability and Tree Growth: A Basin-Wide Analysis," *Global Change Biology* 10 (2004): 704–17, doi:10.1111/j.1529-8817.2003.00772.x.

26 Y. Malhi, J. T. Roberts, R. A. Betts, T. J. Killeen, W. Li, C. A. Nobre, "Climate Change, Deforestation, and the Fate of the Amazon," *Science* 319 (2008): 169–71.

27 D. Nepstad, P. Lefebvre, U. L. da Silva, J. Tomasella, P. Schlesinger, L. Solórzano, P. Moutinho, D. Ray, J. G. Benito, "Amazon Drought and Its Implications for Forest Flammability and Tree Growth: A Basin-Wide Analysis," *Global Change Biology* 10 (2004): 704–17, doi:10.1111/j.1529-8817.2003.00772.x.

28 S. L. Lewis, P. M. Brando, O. L. Phillips, G. M. F. van der Heijden, and D. Nepstad, "The 2010 Amazon Drought," *Science* 331 (2011): 554, doi:10.1126/science.1200807.

29 M. A. Cochrane, "Fire Science for Rainforests," *Nature* 421 (2003): 913–19.

30 S. E. Page, F. Siegert, J. O. Rieley, H. D. V. Boehm, A. Jaya, and S. Limin, "The Amount of Carbon Released from Peat and Forest Fires in Indonesia During 1997," *Nature* 420 (2002): 61–65.

31 M. A. Cochrane, "Fire Science for Rainforests," *Nature* 421 (2003): 913–19.

32 A. R. Holdsworth and C. Uhl, "Fire in Amazonian Selectively Logged Rain Forest and the Potential for Fire Reduction," *Ecological Applications* 7, no. 2 (1997): 713–25.

33 W. F. Laurance, J. L. C. Camargo, R. C. C. Luizão, S. G. Laurance, S. L. Pimm, E. M. Bruna, P. C. Stouffer, G. B. Williamson, J. Benítez-Malvido, H. L. Vasconcelos, K. S. Van Houtan, C. E. Zartman. S. A. Boyle, R. K. Didham, A. Andrade, and T. E. Lovejoy, "The Fate of Amazonian Forest Fragments: A 32-Year Investigation," *Biological Conservation* 144 (2011): 56–67.

34 P. Zelazowsky, Y. Malhi, C. Huntingford, S. Sitch, and J. B. Fisher, "Changes in the Potential Distribution of Humid Tropical Forests on a Warmer Planet," *Philosophical Transactions of the Royal Society* A: 369, no. 1934 (2011): 137–60.

35 P. M. Cox, R. A. Betts, C. D. Jones, S. A. Spall, and I. J. Todderdell, "Acceleration of Global Warming Due to Carbon-Cycle Feedbacks in a Coupled Climate Model," *Nature* 408 (2000): 184–87.

36 C. Huntingford, R. A. Fisher, L. Mercado, B. B. Booth, S. Stich, P. P. Harris, P. M. Cox, C. D. Jones, R. A. Betts, Y. Malhi, G. R. Harris, M. Collins, and P. Moorcroft, "Towards Quantifying Uncertainty in Predictions of Amazon 'Dieback,'" *Philosophical Transactions of the Royal Society* B 363, no. 1498 (2008): 1857–64, doi:10.1098/rstb.2007.0028.

37 Met Office, "Understanding Climate Change Impacts on the Amazon Rainforest," January 2013, http://www.metoffice.gov.uk/research/news/amazon-dieback.

38 J. A. Marengo, C. A. Nobre, G. Sampaio, I. F. Salazar, and L. S. Borma, "Climate Change in the Amazon Basin: Tipping Points, Changes in Extremes, and Impacts on Natural and Human Systems," in *Tropical Rainforest Responses to Climatic Change*, 2nd ed., M. B. Bush, J. R. Flenley, and W. D. Gosling, eds. (Berlin: Springer, 2011), 259–83.

39 R. Fu, L. Yin, W. Lib, P. A. Arias, R. E. Dickinson, L. Huang, S. Chakraborty, K. Fernandes, B. Liebmann, R. Fisher, and R. Myenig, "Increased Dry-Season Length over Southern Amazonia in Recent Decades and Its Implication for Future Climate Projection," *Proceedings of the National Academy of Sciences* 110 (2013): 18110–115, doi:10.1073/pnas.1302584110.

40 D. V. Spracklen, S. R. Arnold, and C. M. Taylor, "Observations of Increased Tropical Rainfall Preceded by Air Passage over Forests," *Nature* 489 (2012): 282–85, doi:10.1038/nature11390.

41 National Research Council, *Abrupt Impacts of Climate Change: Anticipating Surprises*, Pre-publication version (Washington, DC: National Academies Press, 2013).

42 A. Dai, "Increasing Drought under Global Warming in Observations and Models," *Nature Climate Change* 3 (2012): 52–58, doi:10.1038/nclimate1633.

43 M. A. Cochrane, "The Past, Present, and Future Importance of Fires in Tropical Rainforests," in *Tropical Rainforest Responses to Climatic Change*, 2nd ed., M. B. Bush, J. R. Flenley, and W. D. Gosling, eds. (Berlin: Springer, 2011), 213–40.

SPECIALIST'S VIEW: ELLIOTT

1   S. Solomon, D. Qin, M. Manning, Z. Chen, M. Marquis, K. B. Averyt, M. Tignor, and H. L. Miller, eds., *Contribution of Working Group I to the Fourth Assessment Report of the Intergovernmental Panel on Climate Change*, 2007 (Cambridge, UK, and New York: Cambridge University Press, 2007).

2   R. Houghton, "The Emissions of Carbon from Deforestation in the Tropics: Past Trends and Future Potential," *Carbon Management* 4, no. 5 (2013): 539–46.

3   M. Santilli, P. Moutinho, S. Schwartzman, D. Nepstad, L. Curran, and C. Nobre, "Tropical Deforestation and the Kyoto Protocol," *Climate Change* 71 (2005): 267–76.

4   United Nations Framework Convention on Climate Change, "Decision 1/CP.13: Bali Action Plan," in "Report of the Conference of the Parties on its thirteenth session, held in Bali from 3 to 15 December 2007: Addendum: Part Two: Action taken by the Conference of the Parties at its thirteenth session," http://unfccc.int/resource/docs/2007/cop13/eng/06a01.pdf.

5   N. Stern, *Stern Review on the Economics of Climate Change* (London: HM Treasury, 2006).

6   J. Eliasch, *Climate Change: Financing Global Forests* (London: UK Treasury Department, 2008).

7   *The Economist*, "The World's Lungs: Forests and How to Save Them," 25 September–1 October, 2010.

8   *World Resources Institute Blog,* "The REDD+ Decision in Cancun," blog entry by
K. Austin, F. Daviet, and F. Stolle, 20 December, 2010, http://www.wri.org/
blog/2010/12/redd-decision-cancun.

9   A. Angelsen and D. McNeill, "The Evolution of REDD+," in *Analyzing REDD+: Challenges and Choices,* A. Angelsen, M. Brockhaus, W. D. Sunderlin, and L. V. Verchot,
eds. (Bogor, Indonesia: CIFOR, 2012).

10   A. Angelsen, ed., *Moving Ahead with REDD* (Bogor, Indonesia: CIFOR, 2008).

11   A. Larson, "Tenure Matters in REDD+," in *Analyzing REDD+: Challenges and Choices,*
A. Angelsen, M. Brockhaus, W. D. Sunderlin, and L. V. Verchot, eds. (Bogor, Indonesia: CIFOR, 2012).

12   A. Angelsen, S. Brown, C. Loisel, L. Peskett, C. Streck, and D. Zarin, *Reducing Emis-
sions from Deforestation and Forest Degradation (REDD): An Options Assessment Report*
(Washington, DC: Meridian Institute, 2009), http://www.redd-oar.org/links/
REDD-OAR_en.pdf.

13   J. Stoltenberg, "Speech at UN Climate Change Conference in Bali," Norwegian
Government Document Archive, December 2007, http://www.regjeringen.no/
en/archive/Stoltenbergs-2nd-Government/Office-of-the-Prime-Minister/taler-
og-artikler/2007/speech-at-un-climate-conference-in-bali.html?id=493899.

14   A. Angelsen and D. McNeill, "The Evolution of REDD+," in *Analyzing REDD+: Challenges and Choices,* A. Angelsen, M. Brockhaus, W. D. Sunderlin, and L. V. Verchot,
eds. (Bogor, Indonesia: CIFOR, 2012).

15   J. Assunção, C. C. e Gandour, and R. Rocha, *Deforestation Slowdown in the Legal Amazon: Prices or Policies?* (San Francisco: Climate Policy Initiative, 2012).

CHAPTER 9

1   T. Rudel and J. Roper, "The Paths to Rainforest Destruction: Cross-National Patterns of Tropical Deforestation, 1975–1990," *World Development* 25 (1997): 53–65.

2   S. J. Wright and H. C. Muller-Landau, "The Future of Tropical Forest Species,"
Biotropica 38, no. 3 (2006): 287–301.

3   R. S. DeFries, T. Rudel, M. Uriarte, and M. C. Hansen, "Deforestation Driven by
Urban Population Growth and Agricultural Trade in the Twenty-First Century,"
Nature Geoscience 3 (2010): 178–81.

4   S. Hecht and S. Saatchi, "Globalization and Forest Resurgence: Changes in Forest Cover in El Salvador," *Bioscience* 57, no. 8 (2007): 663–72.

5   J. Randers, *2052: A Global Forecast for the Next Forty Years,* A report to the Club
of Rome commemorating the 40th anniversary of *The Limits to Growth* (White
River Junction, Vermont: Chelsea Green Publishing, 2012).

6   C. B. Field, V. R. Barros, D. J. Dokken, K. J. Mach, M. D. Mastrandrea, T. E. Bilir, M. Chatterjee, K. L. Ebi, Y. O. Estrada, R. C. Genova, B. Girma, E. S. Kissel, A. N. Levy, S. MacCracken, P. R. Mastrandrea, and L. L. White, eds., *Climate Change 2014: Impacts, Adaptation, and Vulnerability. Part A: Global and Sectoral Aspects. Contribution of Working Group II to the Fifth Assessment Report of the Intergovernmental Panel on Climate Change* (Cambridge, UK and New York: Cambridge University Press, 2014): Vol I, chapter 4.3.3.1: "Forests and Woodlands."

7   National Research Council, *Abrupt Impacts of Climate Change: Anticipating Surprises*, Pre-publication version (Washington, DC: National Academies Press, 2013).

8   W. F. Laurance, J. Sayer, and K. G. Cassman, "Agricultural Expansion and Its Impacts on Tropical Nature," *Trends in Ecology & Evolution* 29, no. 2 (2013): 107–16.

9   R. S. DeFries, T. Rudel, M. Uriarte, and M. C. Hansen, "Deforestation Driven by Urban Population Growth and Agricultural Trade in the Twenty-First Century," *Nature Geoscience* 3 (2010): 178–81.

10   E. F. Lambin and P. Meyfroidt, "Global Land Use Change, Economic Globalization, and the Looming Land Scarcity," *Proceedings of the National Academy of Sciences* 108 (2011): 3465–72.

11   T. Rudel, "Changing Agents of Deforestation: From State-Initiated to Enterprise Driven Processes, 1970–2000," *Land Use Policy* 24 (2007): 35–41.

12   C. M. Barr and J. A. Sayer, "The Political Economy of Reforestation and Forest Restoration in Asia-Pacific: Critical Issues for REDD+," *Biological Conservation* 154 (2012): 9–19.

13   P. Mayaux, J.-F. Pekel, B. Desclée, F. Donnay, A. Lupi, F. Achard, M. Clerici, C. Bodart, A. Brink, R. Nasi, and A. Belward, "State and Evolution of the African Rainforests between 1990 and 2010," *Philosophical Transactions of the Royal Society B* 368 (2013): 20120300, doi:10-1098/rstb.2012.0300.

14   A. Dai, "Increasing Drought under Global Warming in Observations and Models," *Nature Climate Change* 3 (2012): 52–58, doi:10.1038/nclimate1633.

15   P. Schulte, et al., "The Chicxulub Asteroid Impact and Mass Extinction at the Cretaceous–Paleogene Boundary," *Science* 327 (2010): 1214–1218.

16   Mysterium.com, http://www.mysterium.com/extinction.html.

17   S. Lawson and L. Macfaul, *Illegal Logging and Related Trade: Indicators of the Global Response* (London: Chatham House, 2010).

18   W. F. Laurance, J. L. C. Camargo, R. C. C. Luizão, S. G. Laurance, S. L. Pimm, E. M. Bruna, P. C. Stouffer, G. B. Williamson, J. Benítez-Malvido, H. L. Vasconcelos, K. S. Van Houtan, C. E. Zartman. S. A. Boyle, R. K. Didham, A. Andrade, and T. E. Lovejoy, "The Fate of Amazonian Forest Fragments: A 32-Year Investigation," *Biological Conservation* 144 (2011): 56–67.

19   IUCN, "Global Partnership on Forest Landscape Restoration (GPFLR),"
     http://www.iucn.org/about/work/programmes/forest/fp_our_work/fp_our_
     work_thematic/fp_our_work_flr/more_on_flr/global__partnership_forest_
     landscape_restoration.

20   M. C. Hansen, P. V. Potapov, R. Moore, M. Hancher, S. A. Turubanova, A. Tyu-
     kavina, D. Thau, S. V. Stehman, S. J. Goetz, T. R. Loveland, A. Kommareddy,
     A. Egorov, L. Chini, C. O. Justice, and J. R. G. Townshend, "High-Resolution
     Global Maps of 21st-Century Forest Cover Change," *Science* 342 (2013): 850–53,
     doi:10.1126/science.1244693.

21   J. Clay, "Freeze the Footprint of Food," *Nature* 475 (2011): 287–89.

APPENDIX 1

1   A. Sommer, "Attempt at an Assessment of the World's Tropical Moist Forests,"
    Unasylva 28, no. 112–13 (1976): 5–25.

APPENDIX 2

1   http://www.satimagingcorp.com/characterization-of-satellite-remote-sens-
    ing-systems.html.

2   D. O. Fuller, "Tropical Forest Monitoring and Remote Sensing: A New Era of
    Transparency in Forest Governance?," *Singapore Journal of Tropical Geography* 27
    (2006): 15–29.

3   F. Achard and M. C. Hansen, "Use of Earth Observation Technology to Monitor
    Forests across the Globe" in *Global Forest Monitoring from Earth Observation,* eds. F.
    Achard and M. C. Hansen (Boca Raton, Florida: CRC Press, 2012).

4   http://www.wikipedia.org.

5   NASA and the US Geological Survey developed global data sets from Landsat
    archives, available under the Global Land Survey (GLS) free of cost from the
    Earth Resources Observation and Science (EROS) Center. In 2008 the entire
    Landsat archive was released at no cost to users.

APPENDIX 3

1   F. Achard, H. D. Eva, H.-J. Stibig, P. Mayaux, J. Gallego, T. Richards, and J.-P. Mal-
    ingreau, "Determination of Deforestation Rates of the World's Humid Tropical
    Forests," *Science* 297 (2002): 999–1002.

INDEX

DAVID SUZUKI FOUNDATION

The David Suzuki Foundation works through science and education to protect the diversity of nature and our quality of life, now and for the future.

Our vision is that within a generation, Canadians act on the understanding that we are all interconnected and interdependent with nature. We collaborate with scientists, communities, businesses, academia, government and non-governmental organizations to find solutions for living within the limits of nature.

The Foundation's work is made possible by individual donors across Canada and around the world. We invite you to join us.

For more information, please contact us:

The David Suzuki Foundation
219–2211 West 4th Avenue
Vancouver, BC Canada V6K 4S2
www.davidsuzuki.org
contact@davidsuzuki.org
Tel: 1-800-453-1533

Checks can be made payable to the David Suzuki Foundation.
All donations are tax-deductible.
Canadian charitable registration: (BN) 12775 6716 RR0001
U.S. charitable registration: #94-3204049